Contested Airport Land

Contested Airport Land draws attention to the accelerating airport development in the Global South. Empirical studies provide nuanced analysis of socioeconomic, administrative, and political dynamics on the land beyond the airport grounds, such as the project area of greenfield development, the airport city, or land resources reserved for future airport expansion.

The authors in this book emphasise why airport construction is a politically sensitive issue in low-income and low-middle-income countries, which serve as the last development frontier of the aviation sector. They argue that observed airport development was rather motivated by the perception of airports as engines for national economic growth, while improving air mobility of national populations was not the main driver. Under dominant national development visions, airport-induced dynamics threatened local livelihoods by triggering economies of anticipation, the reconfiguration of land markets, rapid land use changes, a transition from rural to urban livelihoods, the displacement of communities, the perpetuation of human–wildlife conflicts, or inter-ethnic violence. The authors also highlight colonial path dependencies; legal pluralism in land tenure; the hegemonic relations between builders, investors, and the affected residents; as well as strategies of local protest movements.

This book is recommended for readers interested in infrastructure-induced conflicts and environmental injustice.

Irit Ittner is working as a senior researcher in the Programme Environmental Governance at the German Institute of Development and Sustainability in Bonn. Her research interests include unplanned urbanisation, land tenure, social navigation, and processes of transformation in coastal West African and European cities. Irit published on the airport land in Abidjan in *Afrika Focus* (2021), *Urban Forum* (2022), and *Afrique Contemporaine* (2023).

Sneha Sharma works as a junior project manager at the ICON Institut in Cologne after having conducted research at the University of Bonn (2015–2022). Her lived experiences growing up in the busy streets of Kolkata, India, shaped her interest in urban sociology and ethnographic methods. Sneha published *Waste(d) collectors: Politics of urban exclusion in India* (2022). Her work on spatial transformation, affordable housing, and urban renewal in the airport villages of Mumbai were published in *Geoforum* (2023).

Isaac Khambule is a professor of political economy at the University of Johannesburg, South Africa. He was previously an associate professor of political economy at the Wits School of Governance, University of the Witwatersrand, where he taught decision-making in public institutions and worked as a senior lecturer at the University of KwaZulu-Natal. Isaac's research interest is in the relationship between the state, institutions, and development, with a particular focus on the role of the state in economic development and the entrepreneurial state.

Hanna Geschewski is a doctoral researcher in Human Geography at the Chr. Michelsen Institute (CMI) and the University of Bergen in Norway. Her research focuses on the socio-ecological dimensions of displacement and resettlement in South Asia, with a particular interest in human-land relations. Her work on the unfinished airport project in Nijgadh, Nepal, co-authored with M. Islar, was published in the *Journal of Political Ecology* (2022).

Routledge Research in Planning and Urban Design

Routledge Research in Planning and Urban Design is a series of academic monographs for scholars working in these disciplines and the overlaps between them. Building on Routledge's history of academic rigour and cutting-edge research, the series contributes to the rapidly expanding literature in all areas of planning and urban design.

Participatory Spaces under Urban Capitalism
Contesting the Boundaries of Democratic Practices
Markus Holdo

Inclusion and Exclusion of the Urban Poor in Dhaka
Power, Politics and Planning
Rasheda Rawnak Khan

Remodelling to Prepare for Independence
The Philippine Commonwealth, Decolonisation, Cities and Public Works, c. 1935–46
Ian Morley

Co-Creative Placekeeping in Los Angeles
Artists and Communities Working Together
Brettany Shannon, David C. Sloane and Anne Bray

Australia and China Perspectives on Urban Regeneration and Rural Revitalization
Raffaele Pernice and Bing Chen

Contested Airport Land
Social-Spatial Transformation and Environmental Injustice in Asia and Africa
Edited by Irit Ittner, Sneha Sharma, Isaac Khambule and Hanna Geschewski

For more information about this series, please visit: www.routledge.com/Routledge-Research-in-Planning-and-Urban-Design/book-series/RRPUD

Contested Airport Land

Social-Spatial Transformation and
Environmental Injustice in Asia and Africa

Edited by Irit Ittner, Sneha Sharma,
Isaac Khambule and Hanna Geschewski

LONDON AND NEW YORK

Designed cover image: Close spatial entanglement between Mumbai International Airport and informal settlements. Photograph Sneha Sharma

First published 2025
by Routledge
4 Park Square, Milton Park, Abingdon, Oxon OX14 4RN

and by Routledge
605 Third Avenue, New York, NY 10158

Routledge is an imprint of the Taylor & Francis Group, an informa business

© 2025 selection and editorial matter, Irit Ittner, Sneha Sharma, Isaac Khambule and Hanna Geschewski; individual chapters, the contributors

The right of Irit Ittner, Sneha Sharma, Isaac Khambule and Hanna Geschewski to be identified as the authors of the editorial material, and of the authors for their individual chapters, has been asserted in accordance with sections 77 and 78 of the Copyright, Designs and Patents Act 1988.

With the exception of Chapter 1 and Chapter 8, no part of this book may be reprinted or reproduced or utilised in any form or by any electronic, mechanical, or other means, now known or hereafter invented, including photocopying and recording, or in any information storage or retrieval system, without permission in writing from the publishers.

Chapter 1 and Chapter 8 of this book are freely available as a downloadable Open Access PDF at http://www.taylorfrancis.com under a Creative Commons Attribution-ShareAlike (CC-BY-SA) 4.0 license.

Any third party material in this book is not included in the OA Creative Commons license, unless indicated otherwise in a credit line to the material. Please direct any permissions enquiries to the original rightsholder.

Trademark notice: Product or corporate names may be trademarks or registered trademarks, and are used only for identification and explanation without intent to infringe.

British Library Cataloguing-in-Publication Data
A catalogue record for this book is available from the British Library

Library of Congress Cataloging-in-Publication Data
Names: Ittner, Irit, editor.
Title: Contested airport land : social-spatial transformation and environmental injustice in asia and africa / edited by Irit Ittner, Sneha Sharma, Isaac Khambule, Hanna Geschewski.
Description: First published. | New York, NY : Routledge, 2025. | Series: Routledge research in planning and urban design | Includes bibliographical references and index.
Identifiers: LCCN 2024021817 (print) | LCCN 2024021818 (ebook) |
  ISBN 9781032800035 (hardback) | ISBN 9781032800042 (paperback) |
  ISBN 9781003494966 (ebook)
Subjects: LCSH: Airports—Developing countries—Planning.
Classification: LCC TL725.3.P5 C65 2025 (print) | LCC TL725.3.P5 (ebook) |
  DDC 629.13609172/4—dc23/eng/20240718
LC record available at https://lccn.loc.gov/2024021817
LC ebook record available at https://lccn.loc.gov/2024021818

ISBN: 978-1-032-80003-5 (hbk)
ISBN: 978-1-032-80004-2 (pbk)
ISBN: 978-1-003-49496-6 (ebk)

DOI: 10.4324/9781003494966

Typeset in Times New Roman
by Apex CoVantage, LLC

Contents

	List of Figures	ix
	List of Contributors	x
	Foreword	xiii
	ROSE BRIDGER	
1	**Contested airport lands in the Global South**	1
	SNEHA SHARMA, IRIT ITTNER, ISAAC KHAMBULE, SARA MINGORRÍA AND HANNA GESCHEWSKI	
2	**'By now, it feels more like a rumour': navigating the suspended presents and the economy of anticipation for Nepal's Second International Airport**	22
	HANNA GESCHEWSKI	
3	**The rise of infrastructure-induced human–elephant conflict in Sri Lanka: a case study of Mattala Rajapaksa International Airport**	43
	MENUSHA GUNASEKARA AND DISHANI SENARATNE	
4	**A critical review of airport land contestations in India**	62
	SNEHA SHARMA	
5	**Aerotropolis at what cost, to whom? An analysis of the social and economic impacts of the Yogyakarta International Airport, Indonesia**	81
	ELLEN PUTRI EDITA	
6	**The popular appropriation of the airport reserve in Abidjan, Côte d'Ivoire, and strategies to resist displacement**	97
	IRIT ITTNER	

7 **The Durban Aerotropolis: emerging and underlying territorial contestations in South Africa** 120
ISAAC KHAMBULE

8 **Competing aspirations and contestations at Isiolo International Airport, Kenya** 137
EVELYNE ATIENO OWINO AND CLIFFORD COLLINS OMONDI OKWANY

Index *160*

Figures

1.1	Map of airports covered by the case studies.	2
1.2	Spatial model of airport land.	11
2.1	Timeline of events directly related to the SIA project and the relevant historic events in Nepal.	24
2.2	Map of the study area in Bara district, Nepal.	29
2.3	Visible traces of the SIA project in Bara district: (a) project sign along the East–West Highway and (b) SIA project office in Simara.	31
2.4	Online advertisement for real estate in Bara district.	33
3.1	A man passes by a billboard put up by the Airport and Aviation Services Limited featuring the president Mahinda Rajapaksa for accelerating the development of Mattala Airport.	46
4.1	Location of case study sites.	67
5.1	The construction of YIA airport.	82
5.2	Land sales banner along the way to YIA, highlighting vicinity to the airport.	83
5.3	Relocation area of Palihan.	90
6.1	Location of the FHB International Airport.	99
6.2	Freshly reclaimed vegetable beds near the airport wall.	106
6.3	Billboard indicating public landownership.	107
6.4	Gate to the Cité Douane.	109
7.1	Durban Aerotropolis and the areas within one-hour radius.	127
8.1	Map of Isiolo showing the contested boundary along the international airport.	140

Contributors

Rose Bridger is a knowledgeable and acknowledged environmental activist and author of *Plane Truths* (2013).

Ellen Putri Edita is a research associate at the Economic Research Institute for ASEAN (Association of Southeast Asian Nations) and East Asia, based in Jakarta, Indonesia. She is currently involved in research to develop the Regional Knowledge Centre for Marine Plastic Debris, which serves as an information clearing house in ASEAN Plus Three countries. Ellen holds a bachelor's degree in environmental engineering from Diponegoro University, Indonesia, as well as a master's degree in environment and sustainability sciences from Lund University, Sweden. She is passionate about sustainability-related issues, particularly in the complexity of political, social, economy, and environmental relations. Her research interests include environmental governance, political ecology, and waste management.

Hanna Geschewski is a doctoral researcher in Human Geography at the Chr. Michelsen Institute (CMI) and the University of Bergen in Norway. Her research focuses on the socio-ecological dimensions of long-term displacement and resettlement with an emphasis on human-land relations, drawing on fields such as agrarian and peasant studies, refugee and migration studies, and political ecology. Hanna has a background in environmental and sustainability studies and has many years of academic and professional experience in South Asia and the Himalayan region. Her research on the unfinished airport project in Nijgadh, Nepal, co-authored with M. Islar, was published in the *Journal of Political Ecology* (2022).

Menusha Gunasekara is a development professional and an aspiring researcher from Sri Lanka. She co-founded Youth for Action, a volunteer group that is committed to building skills of urban poor children and gender equality. She holds a bachelor's degree in public health from the Asian University for Women, as well as a master's degree in development studies from the University of Sussex, United Kingdom, on a Chevening Scholarship from the United Kingdom's Foreign Commonwealth Development Office. Menusha's research interests include urban poor youth; rural development; labour rights; water, sanitation, and hygiene (WASH); food security; and land use.

Irit Ittner holds a doctoral degree in social anthropology from the University of Cologne, Germany. She worked on natural resources management, on urban development in West Africa, and on ocean governance in Europe. After employment at the University of Bonn (2003–2023), Irit now works as a senior researcher in the Programme Environmental Governance at the German Institute of Development and Sustainability in Bonn. Her research interests include unplanned urbanisation, land tenure, social navigation, and processes of transformation in coastal West African and European cities. Irit published on the airport land in Abidjan in *Afrika Focus* (2021), *Urban Forum* (2022), and *Afrique Contemporaine* (2023).

Isaac Khambule is a professor of political economy at the University of Johannesburg, South Africa. He was previously an associate professor of political economy at the Wits School of Governance, University of the Witwatersrand, where he taught decision-making in public institutions and worked as a senior lecturer at the University of KwaZulu-Natal. Isaac's research interest is in the relationship between the state, institutions, and development, with a particular focus on the role of the state in economic development and the entrepreneurial state.

Sara Mingorría is a Juan la Cierva Incorporation research fellow at the Universitat de Girona, Spain. She is a scholar-activist in the international network Stay Grounded (stay-grounded.org) that works to reduce air traffic in order to build a climate-just transport system. Sara co-founded the platform Zeroport (zeroportbcn.wordpress.com, with more than 160 organisations) for the de-growth of aviation and port infrastructure in Catalonia, Spain, and the feminist research collective FRACTAL (colectivofractal.org). Sara's research focuses on agrarian and environmental conflicts and global changes using the participatory action research approach. She analyses and maps socio-environmental conflicts around the world (Ejatlas.org). Sara also explores the links between feminism, economics, and ecology, as well as bridges between action-research-education and art.

Clifford Collins Omondi Okwany is a research fellow and doctoral candidate at the Department of Political Science at the University of Nairobi, Kenya. His research work is based in Kenya and Somalia, focusing on community-armed groups, radicalisation, countering violent extremism, and ontological security. Clifford was the 2023 peace fellow of the Aland Peace Island Institute in Mariehamn, Finland.

Evelyne Atieno Owino is a research fellow at Bonn International Center for Conflict Studies, Germany. Her research examines the implications of large-scale infrastructure projects to the pastoral communities. She is pursuing a doctoral degree at the University of Bonn. Eveline's work cuts across thematic areas of development, peace, and conflict in Northern Kenya.

Dishani Senaratne is a doctoral researcher at the University of Queensland, focusing on the emergence of Sinhala Buddhist nationalism with the implementation of the 1956 Sinhala-Only Policy. She is also the founder/project director of Writing Doves, a non-profit initiative that employs a literature-based approach to enhance young learners' intercultural understanding. Previously, she taught English at the Sabaragamuwa University of Sri Lanka.

Sneha Sharma holds a doctoral degree in geography from the University of Bonn, Germany. She works as a junior project manager at the ICON Institut in Cologne after having conducted research at the University of Bonn (2015–2022). Her lived experiences growing up in the busy streets of Kolkata, India, shaped her interest in urban sociology and ethnographic methods. Sneha published *Waste(d) collectors: Politics of urban exclusion in India* (2022). Her work on spatial transformation, affordable housing, and urban renewal in the airport villages of Mumbai were published in *Geoforum* (2023).

Foreword

This engaging and informative book, the first on the topic of contested airport land, synthesises a wealth of previously fragmented information. Location-specific chapters, contextualised by insights into the dynamics of the aviation sector, elucidate a diversity of cases of airport land disputes. Interviews with affected people, on-the-ground observations, and literature reviews bring depth and detail. Multi-dimensional tensions over airport land are explored. As well as conflict between authorities and communities threatened with displacement, airport projects can pit communities with competing landownership claims against each other and lead to divisions among authorities pursuing divergent development paths.

The book draws attention to the complexity of airport land contestations and the protracted nature of many disputes. Land can be earmarked, cleared, or fenced off for airport development many years before construction commences. Displacement and loss of livelihoods can begin before evictions and bulldozing of farmland. Real estate speculation triggered by announcement of a project can spark land tensions. There are instances of delayed and mothballed projects stalling other forms of development and inflicting socio-cultural shock.

Communities impacted by airport projects are often disadvantaged, and displacement worsens their situation. Since 2018, I have been part of a small team developing the Map of Airport-Related Injustice and Resistance, a partnership between the Stay Grounded network and the Environmental Justice Atlas (EJatlas) project of the Institute of Environmental Science and Technology – Autonomous University of Barcelona (ICTA-UAB). So far, approximately 120 airport-related cases have been investigated and documented by a wide variety of contributors. The feature map is part of the Global Map of Environmental Justice, comprising 4,000 cases of socioeconomic conflict arising from resource extraction and infrastructure development.

Most of the airport cases on the EJatlas map involve land disputes, ranging in severity from inadequate compensation and rehabilitation to forced eviction, affecting hundreds or thousands of people. In the most serious cases, serious human rights violations and violence resulting in injuries and deaths have occurred. Academic research and reports by civil and environmental organisations provided source material for some of the cases, with others being pieced together from online news and information from non-governmental organisations formed by and to support impacted communities. Cases published on the map represent a fraction of the true scale of the problem. Reaching out through our networks and online research, our team has compiled evidence of at least 300 additional airport-linked land acquisition conflict worth investigating further.

My interest in airport land conflicts began during research for a book, *Plane Truth: Aviation's Real Impact on People and the Environment*. As I gathered snippets of information about destruction of settlements and ecosystems, a pattern of allocation of large land areas, not just for runways and terminals, but encompassing significant areas outside the airport fence, for aviation-dependent commercial and industrial development, often described as an 'aerotropolis' or 'airport city', became evident. My main research resources were online trade magazines, in which aerotropolis development was underpinned by a contradictory narrative: supposedly inevitable emergence of airport cities, as if self-assembled by abstract economic forces, jarred against explanations of the enormity of the requisite physical and regulatory infrastructure. Much-lauded existing airport cities were established on the non-aeronautical revenue model, whereby the airport owns and reaps revenue from land surrounding it.

Airport cities are a recurrent theme in this book, which advances contested airport land as an important field of study. The authors' observations and conclusions offer a helpful framework for future research. The topic is an issue of increasing concern around the world as aviation expansion is enabled by new airports, expansion of existing airports and the emergence of airport cities.

Rose Bridger

1 Contested airport lands in the Global South

Sneha Sharma, Irit Ittner, Isaac Khambule, Sara Mingorría and Hanna Geschewski

About this book

The global aviation industry was growing at a rate of 4% annually in the years before the Covid-19 pandemic and was therefore one of the fastest-growing sources of CO_2 emissions (IPCC 2019). In 2018, only 1% of the world's population was responsible for 50% of global emissions from aviation (Gössling & Humpe 2020), and approximately 80% of the global population has never flown in a commercial aircraft (Gurdus 2017). Aviation is also one of the most unequal industries in terms of freedom of movement and gender equity (Gössling et al. 2019). The sector's growth is based on the construction of new airports and the expansion of existing facilities. Given this context, our book, *Contested Airport Lands*, focuses on socio-political and economic dynamics on the land beyond the fenced airport grounds, such as the project areas of greenfield development, anticipated airport cities, or land resources reserved for future airport expansion. The book assembles eight chapters, including case studies from Nepal, Sri Lanka, India, Indonesia, Côte d'Ivoire, South Africa, and Kenya (Figure 1.1), which describe the manifold consequences of airport development.

The initiative for this book originates from a research project carried out by Irit Ittner and Sneha Sharma in India and Côte d'Ivoire on competing land uses between airport expansion and informal settlements.[1] While conducting the project, the authors found that the inspiring social geographical literature about airports and aviation was disconnected from debates in critical development studies. The literature framed airport conflicts mainly as protests against noise pollution or falling real estate values (due to decreased quality of life around airports; see Boucsein et al. 2017). The sidelining of local knowledge and priorities, threats to local identities by land use changes, inter-generational justice, and aviation's contribution to global warming have also provoked protests against airports (Hornig 2017; Hicks 2022; Mingorría & Conté 2023). Faburel and Levy argued, for example, that 'environmental issues are beginning to structure the future of aeromobility, with nuisance caused by aircraft noise playing a major role in this process' (2009: 211). Researchers, however, have mainly investigated airports in European, North American, Southeast Asian, and Gulf states. Authors have discussed aviation-led economic development but made little reference to development visions

DOI: 10.4324/9781003494966-1

This chapter has been made available under a CC-BY-SA 4.0 license.

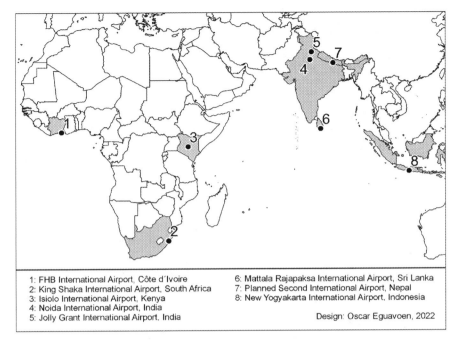

Figure 1.1 Map of airports covered by the case studies.
Source: Map by Eguavoen.

in low- and low-middle-income countries. We argue that the relationship between airports and the inhabitants or local users of airport lands is different in the Global South compared to the Global North because of colonial path dependencies (Button et al. 2019; Chalfin 2019; Tahir 2021) and highly hegemonic relations between builders, investors, and the affected communities. Dynamics differ due to prevalent legal pluralism and insecurity in land tenure, as well as because of greater dependency on foreign funds, weaker accountability, rule of law and democracy, and human rights violations in some countries of the Global South. Unintended consequences of airport construction and new risks include, for example, the perpetuation of human–wildlife conflicts (Gunasekara & Senaratne 2024), emergent land and leadership disputes (Ittner 2023, 2024), growing mistrust towards post-conflict governments (Tahir 2021), or inter-ethnic violence (Owino & Okwany 2024).

Work by scholars, journalists, and activists highlights social and environmental injustice resulting from aviation in both the Global North and the Global South, as well as the contributions of local and global environmental justice movements to stop the growth of the industry. Between 2018 and 2023, the Global Atlas of Environmental Justice (EJAtlas, hereafter), which is a collaborative online mapping platform,[2] recorded 160 cases of protest and mobilisation against new airports or airport expansions worldwide. Researchers and activists like Rose Bridger, who wrote the foreword to this book, have retrieved most of their evidence from local media sources, accounts by local activists and eyewitnesses living and working on the respective airport lands (GAAM 2021a, 2021b; Stay Grounded 2021). An important contribution in this context was the book *Plane Truth* by Bridger (2013), who co-developed the EJAtlas with the ENVJUSTICE project at the University of Barcelona.[3]

In 2021, we invited other scholars via an open call to contribute in-depth studies from the Global South. We got to know each other online and became more familiar with each other's work in the course of the project. Most of the studies existed prior to this book and were realised by scholars originating from the respective countries. Some authors updated their previous datasets for the chapters, or they pooled expertise in new teams of authors. Two, Isaac Khambule and Hanna Geschewski, joined the editorial board to support the finalisation of the manuscript.

The aim of the book is to draw attention to the ongoing acceleration of airport construction and expansion in the Global South, particularly in South Asia, South-East Asia, and Sub-Saharan Africa. South Africa, a middle-income country, does not match the Global South category per se but displays development challenges because land disputes are quite common. Khambule (2024), however, presents a contestation deriving from multi-level governance dynamics and municipal competition (see following text).

The case studies deepen academic understanding of builders' and investors' visions and motivations, as well as of the manifold socioeconomic processes, political dynamics, and contestations accompanying and following airport projects. While three cases each describe the situation in South Asian and Sub-Saharan African countries, one is devoted to a South-East Asian country. Though the geographical distribution of chapter submissions started randomly, the final content indicates the regions where we see the most urgent need for scientific documentation and analysis. Studies about airports as contested infrastructure are rarely found in the first two regions. We also recognise the need to explore and analyse dynamics around new and expanding airports in Latin America (e.g. Lassen & Gallant 2014) but did not receive contributions for this book. In contrast, new airports in South-East Asia have already attracted the attention of regional scholars and stirred academic debates in some countries, such as Indonesia (Kaputra & Putri 2020; Putri & Paskarina 2021; Utami et al. 2021; Heron & Kim 2023; Edita 2024).

Some regional differences in scholarly debate go back to the history of airport construction and expansion, which began earlier in South-East Asia and the Gulf states of the Middle East compared to South Asia and Sub-Saharan Africa (Goetz & Budd 2014b).[4] However, as Njoya and Knowles (2020: 2) underline, 'the commercial aviation industry has become one of the fastest growing sectors in the Global South. . . . For all Global South regions, growth in air traffic has exceeded the world average'. Authors have also pointed out the importance of colonial path dependencies and of huge differences in air transport between countries of the Global South. African countries generally lag behind various indicators, which prompts many African leaders to invest in their infrastructure to address historical socioeconomic and spatial infrastructure inequalities. According to Pirie (2014: 247), 'air transport in Africa is forecast to grow faster than the global average until 2030, and the industry is expected to be a key driver of regional economic development'.

Contributors to this book present nuanced insight into the socioeconomic, administrative, and political dynamics that unfolded before, during, or even decades after airport construction. They provide evidence for the fact that airport construction may be an even more politically sensitive issue in countries at the last development frontier of the aviation sector. Airport development in the Global South is neither necessarily linked to the demand of national industrial sectors nor necessarily targeted improved air mobility of national populations, because most individuals cannot afford flights. As the case studies illustrate, new and larger airports may be driven instead by national development visions, by the growth dynamics of the aviation sector itself, as well as by increased global competition between cities and regions in attracting global investment and international tourism. Air connectivity has evolved into a standard variable in global intra-city competition that motivates further airport development.

This book analyses the complex constellations of land regulation, politics, and strategies that cut across both rural–urban divides and administrative levels and boundaries. They embrace diverse temporalities. The authors have investigated social networks and everyday practices through which social-spatial transformation, political marginalisation, and environmental degradation take place. Repeating themes which the studies bring to the forefront include environmental injustice, the reconfiguration of land markets, the rapid transition from rural to peri-urban and urban livelihoods, multi-level governance dynamics, housing and the displacement of poor and migrant communities, and the dynamic interplay of strategies used by governments, administrative offices, airport authorities, and local protest movements. The hegemonic relation between residents and the developmental state sits at the core of these accounts. The chapters give voice to farmers, pastoralists, and urban dwellers whose livelihoods are or will soon be tremendously affected by new or expanding airports or airport cities. Most residents of the airport lands in this book are economically highly vulnerable. The loss of livelihood resources, such as access to agricultural or pastoral land, water resources, forest, biodiversity, or affordable housing, may be existential for them.

The following chapters show how airport construction is always a political project. Visions of economic development and well-being may differ drastically between populations and political leaders, as well as among different levels and offices of public administration. The cases shed light on the aviation sector in the respective countries, including their historical path dependencies. They present public discourses and national development policies in which aviation plays a prominent role.

The questions connecting the chapters are the following: Who determines how airport land is used now and in the future? Who claims decision-making rights over existing or anticipated airport land? How are these claims justified? What strategies do different interest groups, such as government actors, investors, airport authorities, and the inhabitants of airport lands, pursue to either change or conserve existing land uses? How are airport planning, airport construction, and protests against airports governed? Which socioeconomic and environmental consequences can be observed on the ground? How does multi-level governance play out on airport lands?

Besides drawing attention to the emergent theme of airport-induced conflicts in the Global South and providing rich empirical studies, we aim to synthesise empirical findings. We decided against a general theoretical framework to better grasp the diversity of perspectives. The authors have opted for various concepts, such as legal pluralism, political ecology, environmental justice, urban land governance, political marginalisation, impoverishment risk and reconstruction (IRR), and temporality or territoriality, to explain what is going on around airports.

The different perspectives of the Global North and Global South on aviation

To underline the conditions of our case studies, we argue for the existence of differing perspectives on aviation in countries of the Global North and the Global South. Njoya and Knowles (2020: 1) stated that 'air transport geography has hitherto focused largely on . . . the developed, powerful countries of the Global North, and principally North America, Europe, Japan, Singapore, South Korea, Australia and New Zealand'.

In addition, the socio-cultural and geographical stream of literature has focused on spaces located on airport grounds, such as terminals, airplanes, and runways, and on the airside. Land issues beyond the airport wall have received less scientific attention unless at a wider spatial scale, analysing airports in relation to regional planning, integration, and governance (Knippenberger & Wall 2010).

Other parts of airport research have analysed particular airports, airlines, and their development from a historical, cultural, or economic perspective. Again, most books discuss airports in high-income or high-middle-income countries (for example, Porter 2023). Some of these studies do include discussions of land disputes, displacement, or protest movements (for example, Poulin 2023).

There is also an overrepresentation of Global North literature on the development of the aerotropolis. Kasarda and Lindsay (2011) posited that the transition from airports to airport cities was underpinned by the move away from airports as standard aeronautical spaces to business-oriented and multifunctional projects within and outside them. This is evident in the rise of shopping malls within airports and the number of airports linked to special economic zones in the Global North and in other regions (Nielsen & Da Silva 2017; Wissink 2020).

Airport contestations in countries which we refer to as the Global South (Haug 2020) tend to centre on alternative issues. Residents in our studies, for example, arranged themselves with the noise of the aircrafts and did not phrase it as the major problem,[5] while it was an issue for people living around Mexico City International Airport. They, however, did not succeed in putting the theme on the political agenda and defend themselves against noise (Lassen & Galland 2014).

To provide an example, we found it difficult to identify research on airports in Sub-Saharan Africa that was not business- or management-oriented. Akpoghomeh

(1999) trailblazed with a study on Nigerian aviation. Amankwah-Amoah and Debrah (2013) delivered research on the airline Air Afrique. Njoya (2016) focused on open-sky policies in Africa. Button et al. (2019) assembled contributions, which provided an inventory of the African aviation sector. Njoya and Knowles initiated a first comparative analysis of low- and low-middle-income countries in their special issue of the *Journal of Transport Geography* in 2020. A study by Guntermann (2019) analysed the history and social world of Addis Ababa International Airport and Ethiopian Airways. Kouassi (2021) wrote a history of the international airport in Abidjan and Ivorian aviation.

The situation in South Asia is similar. Hirsh (2016) studied five airports in Asian cities to discuss informal spaces within international air, whereas Nielsen and Da Silva (2017) have written about conflicts arising from a proposed airport in Goa (India).

In countries of the Global South, we identified a particular mix of variables leading to protest movements against airports. Most protest movements do not use argumentation that we know from anti-aviation movements based in the Global North, such as, for example, on the negative impact of aviation on global warming, misled state subsidies, unequal workers' rights, or complaints about noise and pollution (Smith 2019).[6] However, protest movements have taken up these arguments when having aligned with anti-aviation movements in the Global North[7] in order to pool knowledge resources for resistance and to build global networks among airport-opposing movements. Local protests, however, have centred on land disputes due to legal pluralism, unequal political participation, the marginalisation of communities, corruption, human rights violations, or development-induced displacement mirroring protests against other large infrastructure in the Global South. Environmental issues, such as the destruction of habitats and productive landscapes, dust pollution, and water insecurity, have also driven people onto the streets.

For this reason, scholarship that is thematically and conceptually closely related to our case studies emphasises processes of change categorised as development- or infrastructure-induced socioeconomic dynamics with a focus on low- and low-middle-income countries (e.g. Satiroglu & Choi 2017; Johnson et al. 2021; Price & Singer 2019). Infrastructure-induced dynamics may entail the destruction of livelihoods, the impoverishment of resident populations or their displacement, environmental degradation, political oppression, and eventually, violence. Many studies under this scholarship use a political economy or political ecology approach, an environmental justice perspective, or the IRR model proposed by Cernea (1997).

Scholars working on contested infrastructure in the Global South mainly originate from critical development studies, which are multi-disciplinary. Local protest typically sets in before, during, or after the construction of highways, elevated roads, railways, bridges, ports, or large dams (Del Bene et al. 2018; Beier et al. 2022). Because most airports in these regions stem from colonial or post-colonial times and have been operational for decades, studies on contested airports remain rare. This is about to change with a new wave of greenfield development, airport expansion, and aerocity development. Regional development or transport corridors are also debated in the literature, indicating the tendency to plan trans-regional mega-infrastructure made up of several components, including airports. Two chapters in this book reflect on airport development within such corridors (Khambule 2024; Owino & Okwany 2024).

Scholars of contested infrastructure have indicated that infrastructure causes less contestation even though it might damage habitats, livelihoods, communities, or global climate. Generally, populations in low- and low-middle-income countries rather welcome new infrastructure, as it serves as tangible manifestations of progress and modernity and the ability of policymakers to deliver public utilities – even though it might conflict with individual interest. Often, large infrastructure is linked to development promises. Putri and Paskarina point out this aspect based on their study on Yogyakarta International Airport in Indonesia. According to them, it was not the infrastructure that caused the airport conflict but other issues, such as interrupted relations to the land, endangered agricultural income, and fear for the future:

> Although the construction had the same interest, that is, in the name of progress and prosperity – in reality, the conflict still occurred. That means that it was not the progress that became the source of dispute, but there were some other . . . things taken into consideration by the people that eventually some of them put up resistance.
>
> (Putri & Paskarina 2021: 60)

Scholars and activists also emphasise that specific ways of planning and implementing infrastructure have raised disagreement and motivated local opposition. In many projects, there was little or insufficient engagement with affected people. This differs in most countries of the Global North, where public participation schemes are binding for environmental impact assessments. There, statements and questions posed by citizens, environmental organisations and municipalities are systematically collected and archived. Project authorities are obliged to respond. Affected people and communities, therefore, can utilise public participation to communicate and defend their interests. If no public participation schemes are set up, or if project authorities do not respond as required, citizens and activist groups usually use litigation as a powerful tool to improve, delay, or cancel infrastructural projects. The contestation around the new runways at Heathrow Airport in London (Hicks 2022) and at Frankfurt International Airport (Hornig 2017) illustrates such cases.

In the Global South, infrastructure projects are often implemented without integrating them with local livelihoods, many of which are informal businesses (e.g. GAAM 2021a, 2021b; Stay Grounded 2021). Project documentation and planning documents are usually less extensive than in the Global North, even for large infrastructure. Lawsuits against unwanted infrastructural projects are even less common. There is, however, growing evidence that communities and local organisations use litigation in their fights against airports (for example, Lassen & Galland 2014). Modes of project administration and implementation that are frequently reported include misinformation, unkept development promises, irregularities, and elite capture. Studies have also emphasised the relevance of temporal aspects. Anticipation, waiting, and sometimes even the non-materialisation of planned infrastructure may have negative effects on the ground for residents; some reduce over time, while others hinder positive local socioeconomic development for many decades (Geschewski & Islar 2022).

In low- and middle-income countries with high numbers of people living in poverty, governments, planners, and investors envisage airports as engines for economic growth, as competitive hubs of regional and global connectivity. Airports also serve national image building. In the Global North, governments, investors, and builders of airports also communicate these visions – though with much less emphasis on poverty alleviation.

New and larger airports in the Global South – trends in the aviation sector

This section sketches key points in the recent history of the aviation sector to provide readers with contextual knowledge. For decades, airports have been recognised as essential infrastructure for economic development. In countries of the Global South, the idea of catching up with countries of higher income is widespread, as is the perception that catching up would require following European, North American, or Middle Eastern role models.

> Even if a distinction is drawn between economic development and social well-being, growth is still suggested as the one desirable development goal. Those who do not achieve or aim for this goal are considered backward and underdeveloped. This fits in with the understanding of catch-up development under a neoliberal world order. Following the example of the world's top airports, which for a long time were located almost exclusively in the Global North, the aim is to achieve the highest possible growth rates.
> (Guntermann 2019: 21, translated by Ittner)

Initially, international airports were public utilities providing air mobility for passengers, cargo, and the military. That was their core purpose and business. It was at a later point that airports were transformed into profit-making spaces in which air mobility is but one business besides others. The management of many airports changed from states to either public–private partnerships or international corporations with shareholders. In many countries, the state became a shareholder of airports in order to keep some influence over decision-making on aviation. Although international airports are open to the public to use, many are no longer public utilities in the narrow sense but private businesses. Liberalisation encouraged private actors to enter the market by offering a competitive environment. Privatisation was aimed at fostering the efficiency of airline companies by cost-cutting and increasing profitability. In most countries of the Global South, the privatisation of airports began in the late 1990s, when low-income, low-middle-income, and middle-income countries liberalised their economies. From there, the privatisation trend led to leasing out the management and expansion of airports to private firms when the state lacked financial resources. This kind of privatisation allowed the fulfilment of national interests for modernisation and participation in the global competition for foreign investment while the government retained a share of ownership.

With globalisation and the opening up of markets around the globe, there was a pressing demand for integrating additional cities and regions into the global air network. In Sub-Saharan Africa, for example, the air network displayed an orientation towards the centres of former colonial powers in Europe, with a lower number of intra-continental and regional connections (Njoya 2016; Button et al. 2019). Depending on the places of departure and destination, some international flights took ridiculous detours across the continent, via Europe or the Middle East.

The development of cargo and logistics followed the flows of goods across the world as supply chains evolved. Liberalisation and open-sky policies opened up trade restrictions in the 2000s. Airports were seen as key connectors for global growth corridors. The aviation sector was transformed into an industry requiring large capital investments. Building, expanding, and developing airports relied on the participation of multiple stakeholders, particularly in the form of global financial consortia. There were waves of institutional mergers, acquisitions, and restructuring of airlines and these consortia in the following decades due to regional and national liberalisation policies.[8] The active role of national governments in the aviation sector decreased.

In middle-income countries, airports were not only points for taking off and landing. The infrastructure also provided symbolic power to cities and contributed to the expansion of metropolitan areas because they acted as hubs of growth, which was not limited to the airport but extended to ancillary, non-aerial services around the airport. A modern city was defined by the presence of a high-tech international airport which could cater to the needs of commerce and international tourism. Thus, the air mobility of populations was not the main policy target. Civil passengers were typically made up of foreign tourists, international organisations, development cooperation agents (often from the Global North), politicians, international businesspeople, and domestic elites.

Airport development is brought about through planning, governance, and policy changes that facilitate non-aeronautical economic development through airports (Freestone & Wiesel 2016). Experts envisaged an integration of local development with fast-expanding global trade networks to an extent that planners incorporated airports as critical sites for city and regional development. A networked model was seen as ideal, where accommodation, shopping, tourism, luxury services, and cargo freight centres could all be integrated into a cluster-based economy around the airport. Land was at the centre of this model of growth, as proposed by Kasarda and Lindsay (2011), who declared that modern urban development would take place around airports. In Kasarda and Lindsay's 'aerotropolis', the economic development of hinterlands and non-aerial services takes precedence over basic functional needs for connectivity. Airport cities do not serve the aerial global transport networks but are implemented as drivers of local economies. Announcements of new airport construction trigger speculation and create a thriving real estate market where there was none. Airport cities are thus not dependent on aviation but are concentrated hubs that bring together real estate and businesses by hijacking the precincts of airport lands. In the Global South, governments have treated airports as critical projects of modernisation to attract investment, trade, and tourism, thereby prioritising commercial development over the mobility needs of the majority of their populations.

The global expansion of aerial networks is strongly embedded in territorial and place-based processes. Airports are key infrastructure which interlace the global and the local in complex ways that go beyond fulfilling the need for connectivity. Though airports are perceived as complex techno-systems, they are strongly rooted in their locales, given the need for land, security, and labour.

A major shift in African and Asian aviation development has been the intensification of inter-city travel within countries as against the long-distance routes. The use of airports and air mobility significantly increased after the introduction of low-cost carriers, which made flights more affordable for domestic populations. To encourage people from the middle class to travel for business and tourism, point-to-point connecting airports were constructed as against large, hub-based airports.

This book focuses on the territorial aspect of airports. This includes the local political economy, socio-spatial linkages based on the locations of the airport, workplaces, and housing, thus infrastructure such as cargo transport, warehouses, small and medium shops, vegetable production, and housing estates which develop on airport land. (Though workers and inhabitants are possibly connected to the airport through their everyday activities and income strategies, they are usually far less mobile than air passengers and often never even enter the airport.) The main reason for the neglect of socioeconomic dynamics beyond the airport wall in earlier studies, it seems, was the interest in the social world of airports and the airside. Works by Pascoe (2004), Salter (2008), Cwerner et al. (2009), and Adey (2010), for example, suggested social and spatial approaches to studying airports which were novel at that time (aerial life, aeromobilities, the airside of aviation), from which a scholarship evolved during the following years (e.g. Goetz & Budd 2014a; Adey et al. 2017; Baer 2020), one which centred on airports in the Global North.

Contestation around the airport land

Land disputes are central in infrastructural development and social, economic, and environmental change (Chitonge & Harvey 2021; Satiroglu & Choi 2017; Price & Singer 2019; Neef 2023). Land disputes in the context of airport development bring basic political questions to the forefront. Governments operate at multiple, at times competing, levels to plan and implement airports. Laws and environmental regulations define the bearable cost of aviation-induced land use changes. Governance frameworks determine the responsibilities and rights of actors, prescribe project cycles, and conceptualise compensation schemes. Political cultures frame communications with affected residents and influence the political response to project opponents, who defend their stakes on airport land.

Contested airport lands in the Global South 11

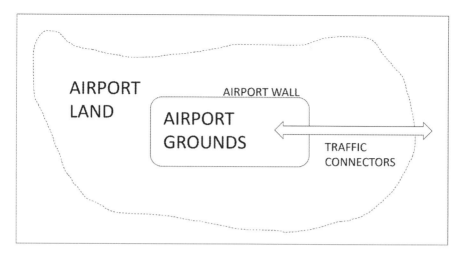

Figure 1.2 Spatial model of airport land.
Source: Design by Ittner.

We conceptualise airport land as the project area of the airport. It consists of the fenced airport grounds, including terminals, runways, other buildings, and aviation-related logistics, as well as areas located beyond the airport wall (Figure 1.2).

Unlike airport grounds, which are indicated by the airport wall or fences, airport land does not usually have tangible borders. The demarcation is rather defined in project documents and laws, as well as by Global Positioning System (GPS) points and lines on maps. In some of our case studies, the airport land is visible on billboards or indicated by the names of hotels or companies. Estate advertisements announce the arrival of the airport and thereby declare plots in the closer or further vicinity to be airport land. Often, airport land can be sensed by the spread of utopian narratives and economic promises. Airport land, as experienced by resident populations and private, small-scale investors, is rather imagined than based on spatial indicators in project documents. Often, there are accelerating land and estate values in these places in relation to the prices of what is not considered airport land. The ownership of plots and houses changes within a short period. Municipalities apply new zoning rules, often in a manner that residents describe as ad hoc and not transparent. Rural land becomes peri-urban or urban by declaration, which entails a number of consequences, including for land value.

Airport roads, metros, and other traffic connectors between airport grounds, airport land, and the surrounding areas are also important for the analysis, as they divide and connect the airport with the city, the countryside, and neighbouring countries. The latter are especially relevant when airports are part of mega-infrastructure, such as regional transportation corridors.

A review of the literature reveals that questions of land rights, ownership, and political authority over land lie at the core of many contestations. This is so because modern international airports require large land resources which already supported various land uses, such as settlements, agriculture, pasture, commerce, or nature conservation, before construction or expansion. New or extended airports always imply massive land use change as well as changes in land tenure and governance. To ensure the security of airport grounds and air traffic, airport authorities prohibit non-aviation uses. Multiple uses are not negotiable. This used to be different, when airports were smaller and their surroundings less intensively used.[9] Nowadays, whoever and whatever is on the land envisaged for new airport facilities will be expropriated in exchange for compensation. In most countries worldwide, the national constitution allows expropriation when important public interests are at stake. Large transport infrastructure, such as airports, counts under this interest.

Project or state authorities conduct inventories of assets, such as houses, fruit trees, cattle, shops, or plantations, and then pay compensation to their owners or, in the case of common pool resources, to communities. From the perspective of the state or airport project, the case of land acquisition and compensation is closed afterwards – but from the point of view of the residents, the process is often not yet finalised due to irregularities and inappropriate compensation payments that do not allow a smooth relocation and recovery.

A refusal to relocate may result in three scenarios: a long delay in airport expansion or construction, with the hope of project cancellation; the forceful eviction of residents (fully or incrementally) to move on with construction; or an avoidance strategy – the construction of satellite airports as greenfield developments. In some cases, authors have observed a mixture of all three scenarios. In Mumbai, for example, debate and resistance to the relocation of the so-called airport slums have been ongoing for decades (Arputham & Patel 2010; Ren 2017). An additional runway could not be built, and part of the air traffic congestion will be channelled to a new greenfield airport in Navi Mumbai. Despite this, airport expansion has been put on hold, leaving the residents of the dense settlements around the Mumbai airport in a permanent limbo of illegality and vulnerability, in which they have developed resistance strategies (Sharma 2024). In Dakar, Senegal, the inner-city airport received a satellite airport, which was built on a rural greenfield site because there was no space reserved for airport expansion. Of course, new land issues and disputes occur on such greenfield sites as well because these are also used by people (Sanogo & Doumbia 2021).

Aviation-induced relocation is framed differently in studies. People or communities refuse to leave because they feel that their ownership rights to the land were (fully or partly) disregarded by authorities. A central variable is the legitimacy of the existing land uses and tenure relations. Legitimacy may be based on diverse factors, such as a long settlement history and resulting customary land rights, as well as customary land transfers, informal land sales, allotments, leases, or rents. The payment of compensation is very tricky in countries where legal pluralism in land tenure persists. In low- and low-middle-income countries, state administrations do not guarantee the full formalisation of land titles because of a limited capacity to manage cadasters and govern land issues. In these countries, several people may hold legal land titles on the same plot, as well as other legitimate temporal or enduring forms of land rights which are not fully recognised by legislation. In such situations, competing claims to land tend to overlap, and land disputes occur. Existing disputes over landownership may turn into open conflicts after large infrastructural projects are announced. Several owners of the same plots may come out and try to formalise the land titles, as the announcement leads to the anticipation of land use change and a revaluation of land resources. If land uses are considered illegitimate or illegal, authorities announce the displacement, evict populations, and demolish buildings without compensation. This may happen, for example, with settlements located on public land. An essential component in such instances is whether inhabitants or other land users hold titles to their plots and buildings and can claim legal ownership.

Airports (and other public infrastructure) in the Global South are often constructed or expanded on communal lands. Communal landownership, however, is not always secured by legal land titles but by a general recognition of customary rights. This turns out to be problematic under rapid land use change scenarios. Migrant communities, meaning people who arrived a century or some decades ago and settled on marginal lands, face the challenge of proving their landownership without holding full citizen rights associated with nationality. Their land use might not have been opposed by local populations because the land was of low value. Swamps are such an example. The legal recognition of migrant land rights in many low- and low-middle-income countries is a long and complicated process which even slows or ends as soon as the state declares an interest in the land. Existing communal land titles have been declared non-valid by authorities in such situations.

Much discomfort, anger, and protest centres on inappropriate or unfair compensation. The literature underlines various reasons for this phenomenon. It seems that planners of relocation and compensation schemes often act without deep knowledge of the historical trajectories of local tenure relations. Land market dynamics on the project land and at the relocation sites are always underestimated. Land disputes or inhabitants' refusal to leave the airport land usually occur because relations between landowners and land users were not sufficiently taken into consideration by compensation schemes. At times, resident landowners and other users receive the same kind of compensation and are put on the same legal footing by authorities. This may entail the upset of local land-giver and land-taker relations, set off local political dynamics, and lead to grievances among owners/users and against the state that endure over many decades (e.g. Ittner 2023, 2024). Often, inhabitants clearly distinguish between the value of their assets and the value of the land they own. While compensation was paid for houses and trees, for example, people were not compensated for the loss of their land. Protest consequently aims at the payment of appropriate compensation and is not necessarily directed against the new airport (Sanogo & Doumbia 2021). Especially when rural communities maintain ritual ties to the land, relocation from ancestral places is felt as an irreversible and painful loss that cannot actually be compensated. In the extreme case of an international airport in Senegal, the communal cemetery was transformed in the building plot for the president's pavilion, an airport building of high symbolic power (Sanogo & Doumbia 2021).

Other disagreements over compensation occur with regard to the categorisation of land and the associated land values for rural and urban land. While greenfield projects begin in rural or peri-urban settings, airports, and especially airport cities, kick off legal land allotment, a construction boom, and rezoning from rural to urban. Inhabitants and land users feel cheated because their compensation payments were based on rural land values, which are much lower than urban land values. They are unable to buy new plots in the vicinity of the airport and watch how other actors gain large profits from land speculation. Compensation is reported as being too low to allow the construction or purchase of other houses. Another concern is the unattractiveness of relocation sites, which may lack connectivity or public infrastructure (Mteki et al. 2017).

Discomfort and disputes have occurred in cases when authorities expropriated and compensated communities but started the construction of the airport some decades later. The current land users and owners do not receive compensation for their losses because earlier generations had received and spent the compensation without yet experiencing the loss. Instead of turning to the elders within their communities, residents confront the state and airport authorities with their frustration. Sometimes, inhabitants learn from the authorities that their grandparents sold their customary land to public authorities many decades ago. They are shown legally binding documents and then have hardly any legal means to fight against their displacement or for compensation.

At times, the affected population exceeds the residents on the airport land. Two studies on the expansion of Julius Nerere International Airport in Tanzania, for example, illustrate how the loss of land intersected with other risks and hardships that were experienced by people who had to leave informal settlements on the airport land and move to a peri-urban area at about a 20 km distance (Mteki et al. 2017; Magembe-Mushi 2018). The resettlement site was peri-urban and less developed than the former living place, with a lack of water, electricity, schools, and health facilities. It also offered less opportunity to earn an income. Displaced households claimed a lack of transparency about compensation payments. Their new neighbours, who were farmers, did not welcome their arrival, because according to Magembe-Mushi (2018), the state had expropriated them in order to create the resettlement site. They had lost their farm and grazing land, partly their homesteads as well, and still awaited compensation payments. Landlessness, joblessness, and homelessness thus affected displaced newcomers and the host community and resulted in social tension and marginalisation, which were accompanied by other negative health and social consequences. This case also included residents' unsuccessful court cases against the government, very long project delays due to lack of finance for compensation and obsolete land legislation over the project period, as well as state oppression against protests by residents. The land plots after relocation were a bit larger than those owned previously, but this could not compensate for the loss of facilities (Mteki et al. 2017).

These dynamics or patterns apply to the construction of other infrastructure, not just airports. Be it roads or dams, infrastructure is planned to represent modernisation and a vision of the future; however, we see that they are, in fact, conduits of power given the role of investment and expert planning which they entail. We argue that our empirical descriptions of and findings on contested airport lands are not extraordinary, even if every case study displays its particularities and is based on extensive data collection in order to show the connections of different variables. They are like any other infrastructure that is used as a symbol of progress and development.

Chapter breakdown

In Chapter 2, Geschewski (2024) considers the case of the suspended 'Second' International Airport in Nijgadh, Nepal. As countries grow and increase their overall economic competitiveness, the need for air connectivity is expected to increase. In the case of Nepal's conceived Second International Airport, it has been 25 years since the airport was planned, yet it has not been delivered by the authorities. Drawing on the political ecology of temporality and anticipation, Geschewski shows how the region of Nijgadh completely changed – moving from agricultural to industrial modes in anticipation of the airport. The speculative nature of the Second International Airport has led to a disruption in local people's lives and investments.

Infrastructure plays a vital part in connecting people to opportunities and is probably one of the most important components of human development. In Chapter 3, Gunasekara and Senaratne (2024) revisit the case of Mattala Rajapaksa International Airport in Sri Lanka, which not only failed to deliver the intended socioeconomic benefits but also contributed to negative social and environmental impact through water-grabbing and disturbing wildlife. The authors consider how this megaproject was grudgingly implemented and overtook the Managed Elephant Reserve development in Hambantota, thus leading to protests. Gunasekara and Senaratne argue that the project was bound to fail because it did not consider the socio-political and ecological impact on local people.

India has embarked on historical and mass infrastructure development that has made the country one of the most globally competitive economies. In its expansion, it was bound to need to expand some of its international airports. As such, Sharma (2024) undertakes a critical review of airport land contestations in India and notes that the country has a long history of land contestations related to airport development. However, Sharma's chapter addresses the lack of studies on community and environmental implications in India through the case studies of Jewa in Uttar Pradesh and Dehradun in Uttarakhand. The author finds that the case of Dehardun was mired in protests to preserve wildlife territory, whereas the case of Jewa revealed the need for fair compensation.

In Chapter 5, on Indonesia's Yogyakarta International Airport, Edita (2024) reveals the political forces driving the decision to establish the airport and airport city. Edita also followed displaced farmers to their relocation site. She analyses the social and economic impacts of airport construction using the IRR model. Airport construction caused landlessness, joblessness, and marginalisation and led to a change in livelihood, which came with the loss of people's identity as farmers.

In Chapter 6, Ittner (2024) explores the popular appropriation of the airport reserve by customary landowners and former land users in Abidjan, Côte d'Ivoire, and illustrates how two spatial forms have evolved as a result: dense informal settlements and peri-urban agriculture with villages. She interweaves the colonial, economic, and migrant history of Abidjan to contextualise the airport city and the ground resistance from those who occupy this space. Authorities find it difficult to demolish buildings after eviction to ensure airport security due to mobilisation and protest against the displacement by urban dwellers. Ittner argues that the living conditions of evicted and returned residents have deteriorated.

Chapter 7 takes the analysis on anticipated airport cities further by using the South African context to study the case of the Durban aerotropolis. Unlike other chapters, which focus on the contestation between authorities and communities, Khambule (2024) shows how this contestation evolved between municipalities. He studied the underlying contestation of the land and benefits of this megaproject between a metropolitan and a local municipality. The author observes that traditional planning mechanisms argue for the metropolis to lead this initiative and that lower-tier municipalities will benefit from trickle-down benefits. However, Khambule argues for shared and balanced growth strategies to evade the emergence of territorial disputes between urban and rural municipalities.

Sub-Saharan Africa has made tremendous progress in infrastructure development after the colonial project. These have been evident in improved human development indices and the connection of many African cities with global aviation networks. In Chapter 8, Owino and Okwany (2024) explore the competing aspirations and contestations at Isiolo International Airport in Kenya. This study is important because it shows the complexities of dealing with land that is not registered within modern constitutionalism. The authors show how the construction of Isiolo International Airport affects the pastoral communities in and adjacent to the airport and raises tensions between the government and affected communities. The land tenure insecurity ensuing from this process creates fragilities that may lead to armed conflict between the Kenyan government and its population.

Findings from this book

To conclude the introductory chapter, we briefly mention the following general conclusions drawn from the empirical case studies. (1) An increase in airport-induced conflict can be assumed for South Asian, South-East Asian, and African countries, as well as in other low- and middle-income countries, with the new wave of airport and airport city construction/expansion. (2) Conflicts over airport lands vary in intensity, ranging from local contestations to region-wide protest movements. Depending on the response by government and airport authorities, conflicts may escalate and display human rights violations, such as illegal imprisonment, gender-specific violence, and state violence against communities. (3) Conflicts over airport land may set in even before construction actually begins, due to the effects of anticipation on local land markets, and may last for many years. (4) Airport-induced conflicts may occur before, during, and long after airport construction or expansion. There may be unintended long-term effects. (5) The interests, policies, and strategies of governments at various levels may strengthen or weaken airport-induced conflicts, especially in the long term. Airports may lead to competition and tension between governments of the same level if airport projects are trans-boundary. (6) Communities or populations threatened by eviction often immigrated into the area decades ago but still lack adequate citizen rights and/or legal land titles to their housing, farming, or pastoral plots. (7) The main forms of protest against airport projects in South Asia, South-East Asia, and Africa include demonstrations, petitions, and the building of political alliances. There are, however, country-specific and cultural forms of resistance used by protest movements that can also be effective.

Acknowledgements

The research project 'Urban villages by the airport: Everyday entanglement of social-economic extremes and negotiations in anticipation of development-induced displacement in Mumbai and Abidjan' was funded by the Fritz Thyssen Foundation (2020–2023, 10.20.2.003EL). This publication was mainly produced at the Geographical Institute of the University of Bonn (GIUB) and was financed by the Open Access Publication Fund of the University of Bonn.

Notes

1 'Urban villages by the airport: Everyday entanglements of social-economic extremes and negotiations in anticipation of development-induced displacement in Mumbai and Abidjan' was conducted at the University of Bonn, Germany.
2 https://ejatlas.org/featured/airport-conflict-around-the-world, 18.09.2023.
3 The initial work was undertaken by Rose Bridger and Sara Mingorría. Each case in the EJAtlas includes information on history and key facts, such as on resistance movements and strategies, institutional support and funding of the airport project, the anticipated and experienced social, environmental, and economic impacts, and the outcome of the protest. The database offers opportunities for the comparative analysis of a large number of case studies, for example, Del Bene et al. (2018), on the resistance to conflictive dams.
4 Goetz and Budd (2014b) provide an idea on how world regions experienced growth in aviation. After Northern America and Europe, they mention rapid growth of the sector in the Asia-Pacific region (Tokyo, Hong Kong, Shanghai, Beijing, Singapore, Seoul, and Sydney), including Thailand. The aviation sector in India and Indonesia started rapid growth by this time, as well as countries in the Middle East (especially the United Arab Emirates and Qatar), Latin America (Brazil, Mexico, Chile, Argentina, Colombia, Peru), and Africa (South Africa, Egypt, Ethiopia, Kenya, Morocco, Nigeria). According to them, Turkey played a special role.
5 Informal settlements in many countries of the Global South are located in the close vicinity of airports.

Permanent noise, for sure, also shows a negative health impact on residents there. In Abidjan and Mumbai, for example, airport noise is not framed as a nuisance but something that one needed to adjust. People told us, for example, how they used the timing of noise from aircraft going to particular (and known) destinations in order to organise their daily routines, such as getting up, sending children to school, or taking a break from work. We admit that air traffic in Abidjan and Mumbai was much lower than in airports covered by Boucsein et al. (2017).

6 A summary of anti-aviation arguments and political suggestions is presented in a brochure by the NGOs Stay Grounded and Kollektiv Periskop: we need a radical de-growth of the aviation sector beyond just sharing the external cost of flying and economic disincentives, as well as policy that supports alternative mobilities (Smith 2019).
7 Relevant initiatives are the Global Anti-Aerotropolis Movement (GAAM) and Stay Grounded.
8 For the EU, 1992 was a milestone year, as the aviation markets of the EU member states were merged into a single aviation market (see Christidis 2016). The open-sky aviation policies mainly emerged in the 2000s for the Global North as well as the Global South, where bilateral and multilateral agreements were made with other countries.
9 Early airports were not always fenced. Air traffic was much less frequent too. Security requirements were fewer. Therefore, historical accounts report on close and even symbiotic relationships between inhabitants and non-aviation users of airport lands and the airports. The airport grounds and runway, for example, could be crossed by residents (Sanogo & Doumbia 2021; Ittner 2023).

References

Adey, P. (2010). *Aerial life. Spaces, mobilities, affects*. West Sussex: John Wiley & Sons.
Adey, P., L. Budd, and P. Hubbard (2017). 'Flying lessons. Exploring the social and cultural geography of air travel'. *Progress in Human Geography* 31(6), 773–791. https://doi.org/10.1177/0309132507083508
Amankwah-Amoah, J. and Y. A. Debrah (2013). 'Air Afrique: The demise of a continental icon'. *Business History* 56(4), 517–546. https://doi.org/10.1080/00076791.2013.809523
Akpoghomeh, O. S. (1999). 'The development of air transportation in Nigeria'. *Journal of Transport Geography* 7, 135–146.
Arputham, J. and S. Patel (2010). 'Recent developments in plans for Dharavi and for the airport slums in Mumbai'. *Environment and Urbanization* 22(2), 501–504. https://doi.org/10.1177/0956247810379936
Baer, H. A. (2020). *Airplanes, the environment, and the human condition*. London: Routledge.
Beier, R., A. Spire, and M. Bridonneau (eds.) (2022). *Urban resettlement in the Global South. Lived experiences of housing and infrastructure between displacement and resettlement*. Abingdon and New York: Routledge.
Boucsein, B., K. Christiaanse, E. Kasioumi, and C. Sewalewski (2017). *The noise landscape. A spatial exploration of airports and cities*. Rotterdam: Nai Publishers.
Bridger, R. (2013). *Plane truth. Aviation's real impact on people and the environment*. London: Pluto Press.
Button, K., G. Martini, and D. Scotti (eds.) (2019). *The economics and political economy of African air transport*. London and New York: Routledge.
Cernea, M. (1997). 'The risks and reconstruction model for resettling displaced populations'. *World Development* 25, 1569–1587.
Chalfin, B. (2019). 'Onshore, offshore Takoradi. Terraqueous urbanism, logistics, and oil governance in Ghana'. *EPD Society and Space* 37(5), 814–832. https://doi.org/10.1177/0263775818800720
Chitonge, H. and R. Harvey (eds.) (2021). *Land tenure challenges in Africa. Confronting the land governance deficit*. Cham: Springer.
Christidis, P. (2016). 'Four shades of open skies: European Union and four main external partners'. *Journal of Transport Geography* 50, 105–114.
Cwerner, S., S. Kesselring, and J. Urry (eds.) (2009). *Aeromobilities*. Abingdon: Routledge.
Del Bene, D., A. Scheidel, and L. Temper (2018). 'More dams, more violence? A global analysis on resistance and repression around conflictive dams through co-produced knowledge'. *Sustainability Science* 13, 617–633.

Edita, E. P. (2024). 'Aerotropolis at what cost, to whom? An analysis of social and economic impacts of the New Yogyakarta International Airport in Indonesia'. In *Contested airport land. Social-spatial transformation and environmental injustice in Asia and Africa*. I. Ittner, S. Sharma, I. Khambule, and H. Geschewski (eds.). Abingdon: Routledge.

Faburel, G. and L. Levy (2009). 'Science, expertise and local knowledge in airport conflicts. Towards a cosmopolitan approach'. In *Aeromobilities*. S. Cwerner, S. Kesselring, and J. Urry (eds.). Abingdon: Routledge, 211–224.

Freestone, R. and I. Wiesel (2016). 'Place making in the rise of the airport city'. In *Place and placelessness revisited*. R. Freestone and E. Liu (eds.). New York/London: Routledge, 168–185.

GAAM (2021a). Aerotropolis. Evictions, ecocide and loss of farmland, part 1. Retrieved 16.01.2024. www.youtube.com/watch?v=nPbXE2YDwVw

GAAM (2021b). Aerotropolis. Evictions, ecocide and loss of farmland, part 2. Retrieved 16.01.2024. www.youtube.com/watch?v=mY96YwLY2Ko

Geschewski, H. (2024). '"By now it feels more like a rumour." Waiting for the Second International Airport in Nepal, suspended presents, and the economy of anticipation'. In *Contested airport land. Social-spatial transformation and environmental injustice in Asia and Africa*. I. Ittner, S. Sharma, I. Khambule, and H. Geschewski (eds.). Abingdon: Routledge.

Geschewski, H. and M. Islar (2022). 'A political ecology of aviation and development. An analysis of relations of power and justice in the (de)construction of Nepal's Second International Airport'. *Journal of Political Ecology* 29(1), 51–75. https://doi.org/10.2458/jpe.2304

Goetz, A. A. and L. Budd (eds.) (2014a). *The geographies of air transport*. Farnham: Ashgate.

Goetz, A. A. and L. Budd (2014b). 'Conclusion'. In *The geographies of air transport*. A. A. Goetz and L. Budd (eds.). Farnham: Ashgate.

Gössling, S., P. Hanna, J. Higham, and D. Hopkins (2019). 'Can we fly less? Evaluating the "necessity" of air travel'. *Journal of Air Transport Management* 81, 101722. https://doi.org/10.1016/j.jairtraman.2019.101722

Gössling, S. and A. Humpe (2020). 'The global scale, distribution and growth of aviation: Implications for climate change'. *Global Environmental Change* 65, 102194. https://doi.org/10.1016/j.gloenvcha.2020.102194

Gunasekara, M. and D. Senaratne (2024). 'The rise of infrastructure-induced human – elephant conflict in Sri Lanka. A case study of Mattala Rajapaksa International Airport'. In *Contested airport land. Social-spatial transformation and environmental injustice in Asia and Africa*. I. Ittner, S. Sharma, I. Khambule, and H. Geschewski (eds.). Abingdon: Routledge.

Guntermann, F. (2019). *Addis Ababa Airport zwischen Globalisierung und Fragmentierung. Ein Hub in Raum und Zeit*. Norderstedt: BoD.

Gurdus, L. (2017). *Boeing CEO: Over 80% of the world has never taken a flight. We're leveraging for growth*. CNBC. https://www.cnbc.com/2017/12/07/boeing-ceo-80-percent-of-people-never-flown-for-us-that-means-growth.html

Haug, S. (2020). 'A thirdspace approach to the "Global South." Insights from the margins of a popular category'. *Third World Quarterly* 42(9), 2018–2038. https://doi.org/10.1080/01436597.2020.1712999

Heron, H. and M. S. Kim (2023). 'Between chili farms and an aerotropolis: The struggle against the new airport in Yogyakarta, Indonesia'. *South East Asia Research* 31(1), 51–71. https://doi.org/10.1080/0967828X.2023.2208371

Hicks, C. (2022). *Expansion rebellion. Using law to fight a runway and save the planet*. Manchester: Manchester University Press.

Hirsh, M. (2016). *Airport urbanism: Infrastructure and mobility in Asia*. Minneapolis: University of Minnesota Press.

Hornig, E.-C. (2017). 'Airport expansion and public protest. The democratic dilemma of vertically asymmetric policies'. *European Policy Analysis* 3(2), 324–342.

IPCC (2019). *Climate change and land: An IPCC special report on climate change, desertification, land degradation, sustainable land management, food security, and greenhouse gas fluxe in terrestrial ecosystems (SRCCL)*. IPCC.

Ittner, I. (2023). 'Emergent disputes over land and leadership in urban villages on the airport reserve in Abidjan'. *Afrique Contemporaine* 276, 175–201. https://doi.org/10.3917/afco1.276.0175

Ittner, I. (2024). 'The popular appropriation of the airport reserve in Abidjan, Côte d'Ivoire, and strategies to resist displacement'. In *Contested airport land. Social-spatial transformation and environmental injustice in Asia and Africa*, I. Ittner, S. Sharma, I. Khambule, and H. Geschewski (eds.). Abingdon: Routledge.

Johnson, C., G. Jain, and A. Lavell (eds.) (2021). *Rethinking urban risk and resettlement in the Global South*. London: UCL Press.

Kaputra, I. and P. W. Putri (2020). 'The precarity of peri-urban resistance. A resistance to the forced eviction of Pasar VI village and the development of Kualanamu International Airport, North Sumatera'. *PCD Journal* 8(1), 49–67. https://doi.org/10.22146/pcd.v8i1.419

Kasarda, J. D. and G. Lindsay (2011). *Aerotropolis. The way we will live next*. London: Allan Lane/Penguin Books.

Khambule, I. (2024). 'The Durban Aerotropolis. Emerging and underlying territorial contestations in South Africa'. In *Contested airport land. Social-spatial transformation and environmental injustice in Asia and Africa*. I. Ittner, S. Sharma, I. Khambule, and H. Geschewski (eds.). Abington: Routledge.

Knippenberger, U. and A. Wall (eds.) (2010). *Airports in cities and regions. Research and practice*. 1st International Colloquium on Airports and Spatial Development. Karlsruhe: KIT Scientific Publishing. https://doi.org/10.5445/KPS/1000017332

Kouassi, K. K. S. (2021). *Avenir du transport aérien international en Cote d'Ivoire à l'horizon 2065*. Thèse de géographie pour l'obtention du grade de docteur. Université Félix Houphouët-Boigny, Abidjan.

Lassen, C. and D. Galland (2014). 'The dark side of aeromobilities. Unplanned airport planning in Mexico City'. *International Planning Studies* 19(2), 132–153.

Magembe-Mushi, D. L. (2018). 'Impoverishment risks in DIDR in Dar es Salaam City. The case of airport expansion project'. *Current Urban Studies* 6, 433–454.

Mingorría, S. and A. Conté (2023). 'No la ampliación del aeropunto El Prat'. *Papers Regió Metropolitana de Barcelona: Territori, estratègies, planejament* 65, 117–131.

Mteki, N., T. Murayama, and S. Nishikizawa (2017). 'Social impacts induced by a development project in Tanzania. A case of airport expansion'. *Impact Assessment and Project Appraisal* 35(4), 272–283.

Neef, A. (2023). *Tourism, land grabs and displacement. The darker side of the feel-good industry*. London/New York, NY: Routledge.

Nielsen, K. B. and S. J. S. Da Silva (2017). 'Golden or green? Growth infrastructures and resistance in Goa'. In: *Urban utopias. Palgrave studies in urban anthropology*. T. Kuldova and M. Varghese (eds.). Cham: Palgrave Macmillan, 53–73.

Njoya, E. T. (2016). 'Africa's single aviation market. The progress so far'. *Journal of Transport Geography* 50, 4–11.

Njoya, E. T. and R. D. Knowles (2020). 'Introduction to the special issue: Air transport in the Global South'. *Journal of Transport Geography* 87.

Owino, E. A. and C. C. O. Okwany (2024). 'Competing aspirations and contestations at the Isiolo International Airport, Kenya'. In *Contested airport land. Social-spatial transformation and environmental injustice in Asia and Africa*. I. Ittner, S. Sharma, I. Khambule, and H. Geschewski (eds.). Abingdon: Routledge.

Pascoe, D. (2004). *Airspaces*. London: Reaktion Books.

Pirie, G. (2014). 'Geographies of air transport in Africa. Aviation's "last frontier"'. In *The geographies of air transportation*. A. A. Goetz and L. Budd (eds.). Abingdon/New York, NY: Routledge, 247–266.

Porter, E. (2023). *A people's history of SFO. The making of the Bay Area and an airport*. Oakland, CA: University of California Press.

Poulin, É. G. (2023). 'Mirabel Airport. In the name of development, modernity, and Canadian unity'. *Economic Anthropology* 10(1), 10–31.

Price, S. and J. Singer (2019). *Country frameworks for development displacement and resettlement. Reducing risk, building resilience*. Abingdon/New York, NY: Routledge.

Putri, D. N. C. and C. Paskarina (2021). 'Strategies to de-escalate the conflict of Yogyakarta International Airport Construction'. *Mimbar* 37(1), 57–66.

Ren, X. (2017). 'Fragile entrepreneurialism: The Mumbai Airport slum redevelopment project'. In *Entrepreneurial urbanism in India. The politics of spatial restructuring and local contestation*. S. K. Chandrashakar (ed.). Singapore: Springer, 157–173. https://doi.org/10.1007/978-981-10-2236-4_8

Salter, M. B. (2008). 'Airport assemblage'. In *Politics at the airport*. M. B. Salter (ed.). Minneapolis: University of Minnesota Press, IX–XIX.

Sanogo, A. and L. Doumbia (2021). 'Land tenure and public infrastructure. Airport building in Côte d'Ivoire and Senegal'. In *Land tenure challenges in Africa. Confronting the land governance deficit*. H. Chitonge and R. Harvey (eds.). Cham: Springer, 203–225.

Satiroglu, I. and N. Choi (2017). *Development-induced displacement and resettlement. New perspectives on persisting problems*. London/New York, NY: Routledge.

Sharma, S. (2024). 'Survival and confrontation. Housing repair practices on Mumbai's airport land'. *Geoforum* 148, 103934.

Smith, T. (ed.) (2019). *Degrowth of aviation. Reducing air travel in a just way*. Vienna: Stay Grounded/Kollektiv Periskop.

Stay Grounded (2021). Webinar Airport conflicts. Struggles for environmental justice. Retrieved 16.01.2024. www.youtube.com/watch?v=WDL6q2ZkEHE

Tahir, A. I. (2021). 'Critical infrastructures as sites of conflict over state legitimacy: The case of Hargeisa Airport in Somaliland, Northern Somalia'. *Geoforum* 125, 110–119. https://doi.org/10.1016/j.geoforum.2021.06.019

Utami, W., D. Nurcahyanto, and S. Sudibyanung (2021). 'Economic impact of land acquisition for Yogyakarta International Airport'. *Mimbar* 37(1), 150–160.

Wissink, H. (2020). 'Evaluating the aerotropolis model for African cities. The case of the Durban aerotropolis'. In *Reflections on African cities in transition. Selected continental experiences*. P. S. Reddy and H. Wissink (eds.). Cham: Springer, 183–211.

2 'By now, it feels more like a rumour': navigating the suspended presents and the economy of anticipation for Nepal's Second International Airport

Hanna Geschewski

Introduction

Arriving in the small town of Nijgadh, in Nepal's Terai region, you will notice that one of the few commercial accommodations along the main road is the Tara Inn Jungle Safari and Airport Hotel. For someone unfamiliar with the area, the name may seem odd because an airport is nowhere in sight. However, the hotel is one of several indications that planes have indeed been expected in the skies over Nijgadh.

Around 25 years ago, an 80 km² area in the neighbouring municipality of Kolhabi, a short rickshaw ride from Nijgadh, was selected as the site for Nepal's Second International Airport (SIA). Thus far, only some preliminary planning and site preparation work have been completed, but the lives of the local population have since been shaped by its anticipated arrival. Aspirations for an airport-induced surge in regional economic activity and tourism have already translated into visible transformations, such as the conversion of agricultural land into building plots, fuelled by rising land prices, and the emergence of new businesses and infrastructure, such as the Tara Inn Hotel and the new expressway connecting the airport with Kathmandu, which is currently under construction. However, the airport project has also left more subtle marks, for example, on people living within the construction area, who have remained in constant uncertainty about their resettlement for years.

In this chapter, I investigate how the anticipated presence of the airport, despite its present absence, has affected communities and their socio-environment since the project was first announced in the early 1990s. Drawing on concepts of temporality and anticipation from political ecology, as well as empirical research on unfinished and imagined infrastructures, I show the multidimensional impact that infrastructure can have long before it is physically realised (Carse & Kneas 2019; Schindler et al. 2019; Braun 2020).

Studying emerging infrastructures through an anticipatory lens allows for an understanding of 'how things come to matter before they *become* "matter"' (Haines 2018: 397). This 'anticipatory mode' brings the less-tangible aspects of socio-ecological change into focus and highlights the political significance of promise and anticipation in development initiatives, which may have very tangible consequences (Adams et al. 2009: 248).

While any infrastructure in the making comes with certain hopes and fears, anticipation, in the case of the Second International Airport, is marked by its

DOI: 10.4324/9781003494966-2

substantial delay. Halted by years of political upheaval, investor withdrawal, and continued Supreme Court stay orders over potential environmental damage, the airport remains stuck in the planning phase, and its completion is increasingly in question (06.10.2020, *The Kathmandu Post*; 03.02.2021, *The Kathmandu Post*).

It is this process of infrastructural protraction, in which waiting becomes the norm, that Carse and Kneas define as the 'suspended present', a concept that highlights the power of suspension in 'rework[ing] experiences of past, present, and future' (Carse & Kneas 2019: 18). Significant deviations from the original project timeline are a common empirical phenomenon in large-scale development initiatives, and increasing attention is being paid to the impact of delays on socioeconomic, environmental, and political processes (Delgado 2018; Carse & Kneas 2019).

Engaging with the temporalities of the airport project – its convoluted history, ongoing delay, and uncertain future – draws attention to the meaning of time in the project. While politicians and project officials use the power of anticipation to paint promising futures, they disregard the impact of the project's years-long suspension on local residents. While landowners have benefited from rising property values fuelled by the 'soon to come', environmental activists have used delay, that is, actively prolonging the 'yet to come', as an effective means of resistance.

The aim of this chapter is to decipher the dynamic implications of the unfinished airport project for proximate communities and landscapes, using temporality as an overarching framework. The analysis is guided by two questions: (1) How has the Second International Airport (SIA) project, despite its physical absence, already influenced local communities and landscapes, and (2) what are the material and non-material impacts of the prolonged suspension of the airport construction on local communities and landscapes?

In what follows, a background section provides an overview of events related to the airport up to the present, followed by a conceptual section on temporalities in political ecology, a short methodological section outlining the applied approach of the fieldwork, and an empirical and discussion section. Finally, there is a concluding section.

Background

While the airport near Nijgadh is often presented as Nepal's latest megaproject, plans for its construction have been in place for more than two decades. The following section provides an overview of the history of the project and the controversies surrounding it. A timeline of relevant events can be found in Figure 2.1.

Since the first aircraft landed on Nepali territory in 1949, air transport has played an important role in connecting the landlocked country to the outside world, boosting tourism, and facilitating regional mobility across its difficult terrain. However, Nepal's aviation history has also been marked by catastrophic plane crashes, increasing congestion, and slow technological progress (Harris 2021). Changing national governments have repeatedly attempted to modernise the country's aviation sector, thus far with disputable success.

Already in its Sixth Five-Year Plan from 1980 to 1985, the government of Nepal, under then–prime minister Surya Bahadur Thapa, had envisaged the search for an

24 Hanna Geschewski

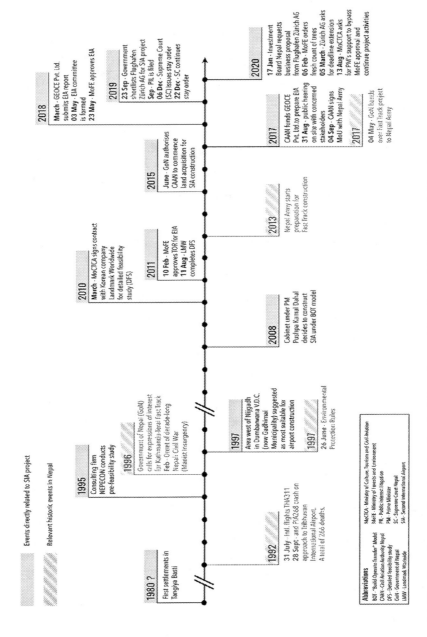

Figure 2.1 Timeline of events directly related to the SIA project and the relevant historic events in Nepal.
Source: Geschewski (2021).

'alternative to Tribhuvan International Airport' (National Planning Commission 1980: 55). To date, the country's only international airport, in Kathmandu, accommodates inbound and outbound international and domestic flights on a single runway, handling up to 7.3 million passengers a year (in pre-Covid-19 times), but its expandability is limited by topographical constraints (CAAN 2020; 13 February 2020, *The Record*).

In 1995, the government, under then–prime minister Sher Bahadur Deuba, commissioned the consulting firm Nepal Engineering Consulting Services Centre Limited (NEPECON) to conduct a feasibility analysis of eight potential locations for a Second International Airport, which deemed the current project site near Nijgadh the most suitable (Shah 2019). The selection of the densely forested area was officially based on factors such as land availability, space for potential future expansion, good connectivity to major cities and the Indian border, and topography (13.02.2020 *The Record*). Critics of the project have since questioned this rationale, as explained later in the text. Shortly afterward, in 1996, the Government invited proposals for the construction of the 75-km-long Kathmandu-Terai Expressway, reducing travel time from the capital to Nijgadh from five hours to one hour (Khatiwada & Aryal 2019; Shakya 2019). However, years of political turmoil during and after Nepal's decadelong Civil War beginning in 1996[1] stalled both projects.

Although the Tenth Five-Year Plan of 2002 mentioned the 'construction of a well-facilitated international airport in the Terai area', it took until 2010 for the Government, under then-Prime-Minister Madhav Kumar Nepal, to commission the Korean company LandMark Worldwide to conduct a detailed feasibility study for the Second International Airport (National Planning Commission 2002: 231; 16.07.2019 The Record). At the time, the company retained the option of implementing the project as part of a build-own-operate-transfer (BOOT) model in addition to conducting the study.

After negotiations with LandMark failed, the Ministry of Culture, Tourism, and Civil Aviation, as the responsible implementing authority, had to re-tender the project and finally shortlisted Flughafen Zurich AG in 2019 (Investment Board Nepal 2019; 24 September 2019, *The Himalayan Times*). However, the government's hopes of bringing in an international investor were dashed once again when Zurich AG pulled out in September 2020, officially citing a change in the company's investment priorities related to the Covid-19 crisis (23.09.2020 *Handelszeitung*).

Following Zurich's withdrawal, the government has repeatedly pledged to implement the construction plans, if necessary, even with state funding (03.02.2021 *The Kathmandu Post*). While presenting the state budget in May 2021, however, Finance Minister Bishnu Poudel announced that funds would only be allocated for completing preparatory work and selecting an international investor (29.05.2021 *Aviation Nepal*; 30.05.2021 *My Republica*). This leaves the financing of the airport project uncertain.

Since plans for the construction of the SIA gained momentum, the project has been accompanied by controversy. While the project is touted by government officials as a boost to the regional and national economies, its purpose, size, and adverse environmental impacts have frequently been questioned (19.12.2019 *The Kathmandu Post*).

Upon completion, the airport, estimated to cost EUR 3 billion, would handle up to 15 million passengers a year, making it the largest air transport hub in South Asia (Investment Board Nepal 2019; 19 December 2019, *The Kathmandu Post*). However, given the rapid expansion of aviation infrastructure in neighbouring China and India (see chapter by Sharma in this edition), the critics of the airport have expressed doubts that the SIA will ever reach its planned capacity (Bhattarai & Conway 2021). In addition, two other international airports are nearing completion, one in Nepal's second largest city, Pokhara, and one in Bhairawa, near the Indian border. While critics believe that these two projects will make the airport in Nijgadh redundant, government officials have argued that the Pokhara Regional International Airport (PKR) and the Gautam Buddha Airport (BWA), in Bhairawa, will mainly function as regional hubs, because they can only handle medium-haul aircraft and are too far from Kathmandu to serve as a direct alternative to Tribhuvan International Airport (17.08.2020 *OnlineKhabar*; 01.06.2018 *Nepali Times*).

Further backlash emerged following the release of the project's environmental impact assessment (EIA) in 2018. The assessment, conducted by Nepali engineering firm GEOCE Consultants (P) Ltd. and approved by the Ministry of Forests and Environment, revealed that up to 2.4 million trees would have to be felled to clear the project site (GEOCE 2018; Shah 2019). The area serves as an elephant migration corridor and a buffer zone for the nearby Chitwan and Parsa National Parks (Thakur 2015). According to forest experts, deforestation could significantly impact regional ecosystems and biodiversity, disrupt local groundwater levels, and lead to increased human–wildlife conflicts (compare to the chapter on Sri Lanka by Gunasekara and Senaratne in this edition) (Chernaik 2019; NEFEJ 2019). The government has since withdrawn its original plans, claiming that, initially, only about 330,000 trees would have to be felled and 25 saplings per tree would be planted as compensation (16.07.2019 *The Record*; Shah 2019). However, the planning authorities have yet to specify the total environmental damage of all project phases and formalise a mitigation plan.

Project opponents also pointed to technical errors in the EIA report and alleged that the selection of the area near Nijgadh was influenced by political agendas and alternative sites were not sufficiently considered (Shah 2019; 13 February 2020, *The Record*). Section 68 of the 1993 Forest Act lists the complete lack of alternative options as one of only three conditions under which forests may be cleared (28.03.2019 *The Record*). The long list of criticisms led to two public interest litigations, in which groups of environmental activists, lawyers, and forest and wildlife experts demanded the examination of alternative sites and the preparation of a new EIA. In response, the Supreme Court of Nepal ordered, in December 2019, that construction be halted until the extent of the project's impact on the local environment has been properly assessed (06.10.2020 *The Kathmandu Post*).

In May 2022, the Supreme Court reaffirmed its prior ruling, delivering a verdict that prohibits the government and other project stakeholders from proceeding with airport construction (27.05.2022 *The Kathmandu Post*). The court directed the involved parties to establish an expert group tasked with a reassessment of the project, including an exploration of alternative locations (10.11.2023 OnlineKhabar). Following the court's order, a group of government-assigned experts was formed, which ultimately (and maybe unsurprisingly) concluded that Nijgadh remains the most suitable site for airport development (09.11.2023 *My Republica*). Ever since, government officials have repeatedly expressed their commitment to the airport's construction, also evident in its recurring inclusion in annual budgets and the allocation of symbolic funds, such as the EUR 2.4 million designated in the 2022–2023 fiscal year (12.06.2023 OnlineKhabar).

Nevertheless, the absence of a confirmed funding source for the required EUR 3 billion (EUR 1.2 billion for the first phase) leaves the next course of action undetermined.

Conceptual framework: temporality of unfinished infrastructures

I had first come across the Nepali airport project in 2019, in the wake of the public outcry over its projected environmental impact, but it was not until my fieldwork in the Nijgadh region that I realised that its impact had begun to unfold long ago. What was, for some, a looming future had, for others, already shaped their past and present.

Throughout my research, time and temporal motives were recurring themes. I noticed how local residents navigated notions of nostalgia and anticipation, how promises of a better future were used by politicians to justify decisions, and various actors differently experienced and portrayed waiting and deferral. I was intrigued to review my data for this chapter using a temporal framework, specifically one concerning the temporality of *unfinished* infrastructures.

Temporality, as a framework for the study of infrastructure projects, has gained increasing attention in the social sciences, signalling a shift away from a sole focus on spatiality (Hetherington 2014; Harris 2021). Recognising that infrastructure unfolds not only in space but also in time ensures that its 'technical and logistical sides . . . are not privileged over, or seen as separate from, its social and political' (Appel et al. 2018).

Rather than focusing on a project's final impacts, a temporal lens helps identify its 'dynamic series of consequences over time', such as 'the production of social meaning and culture, processes of capital accumulation and political strategies' (Braun 2020: 854; Fent & Kojola 2020: 821). Social anthropologists have mainly focused on the first aspect, analysing how emerging infrastructures have shaped people's everyday lives and perceptions of the past, present, and future (Delgado 2018; Haines 2018; Rest 2019). Political ecologists have concentrated on the political significance of time in infrastructure and on the strategic use of temporal narratives, for example, about promising futures (Braun 2020; Fent & Kojola 2020).

A common realisation is that infrastructure projects extend over complex timelines beyond the linearity of planning, construction, and operation (Appel et al. 2018). Such timelines are often marked by disruptions caused by budget shortfalls, political shifts, or societal resistance, resulting in projects remaining unfinished for an indefinite period (Carse & Kneas 2019). The Second International Airport in Nepal, launched over 25 years ago, is just one example of a larger empirical phenomenon of megaprojects that have deviated sharply from their original schedules, also known as 'white elephant projects' (Okebugwu & Omajeh 2014: 1).

What unfinished projects have in common is their potential to trigger change and 'reshape social and political life', even if – or precisely because – they have not yet been fully implemented (Carse & Kneas 2019: 17; Braun 2020). Studies that take the material incompleteness of infrastructures as a vantage point show that a project's absence or presence goes beyond the physical and challenge common perceptions of how and from what point in time a project comes into 'existence'.

In an effort to categorise the phenomenon of unfinished infrastructures, Carse and Kneas (2019: 18) coined the concept of 'suspended presents'. This heuristic refers to projects that are significantly delayed while retaining the possibility of completion. The concept, for one, helps explore the affective states that project suspension can evoke among proximate communities. Anticipation of what is to come, elicited by promises of prosperity, can be overshadowed by what is not there; notions of hope may turn to disillusionment, fear, and rejection. In this way, suspension shapes people's attitudes toward the project and the authorities, potentially jeopardising their support and provoking resistance.

However, suspension also has more material effects. It can visibly shape landscapes by marking off large areas as indefinite construction sites (Delgado 2018). It can also delay decisions on the resettlement of people, putting them in a limbo, and hold up local development because of fears that investments will become redundant after project completion (Kirchherr et al. 2016; Rest 2019). However, suspension can also enable alternative visions of development and counter-movements because it challenges the inevitability of a project (Carse & Kneas 2019).

Suspended infrastructures disclose a disjuncture between 'a promise of progress and an unfinished reality' (Delgado 2018: 60). However, anticipation can be a powerful tool with which to conceal this discrepancy through narratives of future benefits. Cross calls this strategic use of the promising futures of infrastructures the 'economy of anticipation' (Cross 2015: 424). This concept illustrates how anticipation 'shapes relations of power and consent' by maintaining people's support for the project despite persistent delays, as well as by dismissing complaints (Cross 2015: 426). In this way, anticipation becomes a powerful driving force in securing capital flows and speculative investments because it fuels people's hopes of economic gains from the project.

Emphasising the temporal dimensions of the Nepali airport project by using concepts of 'suspended presents' and the 'economy of anticipation' allows for a nuanced understanding of its multi-layered consequences along entangled project timelines. As Haines notes, an anticipatory lens helps elucidate 'how promises of "development" bear cultural, sensory, and material effects for people living in [the] midst' of an unfinished project (Haines 2018: 398). After the conceptual basis has been presented, the next section provides information on the study area in Bara district and briefly outlines the methodological approach.

Study site and methods

The findings presented here draw mainly from empirical data collected during eight weeks of fieldwork in the Nijgadh area in early 2020.

Study site

The SIA project is to be implemented on an area of 80 km² in Bara district, Province 2. The site is bounded by the Pasaha River to the west, the Lal Bakaiya River to the east, and the East–West Highway to the north (see Figure 2.2). The nearest towns are Nijgadh and Simara, where the SIA project office is located. Nearly 95% of the project area is covered by a dense sal tree forest (*Shorea robusta*), providing a habitat for over 500 bird species and 22 endangered wildlife species (Shah 2019; 16 July 2019, *The Record*). Together with the Parsa and Chitwan National Parks, the forest forms a major wildlife migration corridor (NEFEJ 2019).

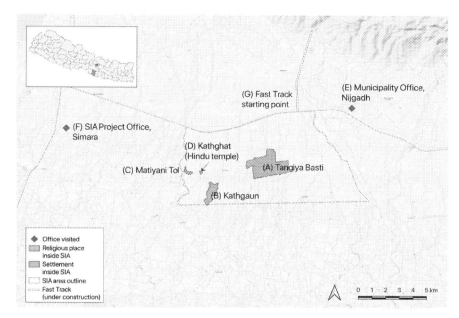

Figure 2.2 Map of the study area in Bara district, Nepal.
Source: Geschewski (2021).

The remaining 5% of the project site includes three settlements whose residents face eviction in the course of SIA: Tangiya Basti, Kathgaun, and Matiyani Tol (Shah 2019).

Tangiya Basti is the largest settlement, with almost 1,500 households. The first few hundred people moved here in the 1970s as tree planters under the government's *taungya*[2] afforestation programme. Many were members of the Tamang and Magar ethnolinguistic groups, hailing from the hilly regions of Nepal, who had previously lost their land to floods and landslides. To date, no one in Tangiya Basti owns the land they have lived on for over four decades (18.07.2019 *The Record*). Although they were originally promised land titles for 0.01 km² per household by the government, these were never granted, and after the end of the *taungya* programme around 1990, half of the planters were relocated to Kailali, leaving the remaining families on their own and without the means to move elsewhere (19.08.2019 *The Record*; Geschewski & Islar 2022).

About 3 km west of Tangiya Basti along the forest line is Kathgaun, the second largest settlement. People began to settle here about 300 years ago (22.07.2020 *The Record*). Most inhabitants belong to the *Tharu*, an indigenous group of the Terai region, and all but 16 of the 132 households used to own their land. However, the government commenced a land acquisition process for the airport in 2015, and by 2017, most of the 112 households eligible for compensation had handed over their land (Bridger 2017). Only a handful of households still refuse to sell.

The smallest settlement, Matiyani Tol, is located directly on the banks of the Pasaha River, west of the much-frequented Hindu Kathgath Temple. The village was founded in the early 2000s by landless Dalit families from other parts of Terai (22.07.2020 *The Record*). None of the 40 households has landownership documents.

Data collection and analysis

My primary data largely draws from observations at various sites in the Nijgadh Region and from interviews with various interest groups in the airport project.

I spent several weeks in the three affected settlements, where I participated in numerous informal conversations and took several narrative walks through the villages and the forest, documenting my impressions through photos and videos. In addition, I visited the towns of Simara and Nijgadh and their surroundings, looking for signs of the airport and talking to Nijgadh residents about the airport and its implications for Nijgadh. Another key source of data were 23 semi-structured and open interviews, including five group interviews, which I conducted with residents of the three affected settlements; local government representatives; project authorities; and environmental activists, scientists, and legal experts in Kathmandu.

I supplemented my primary data with project documents, press releases, legal and policy documents, and accounts from print and social media. In the following section, I present the results of my interview, document, and media content analysis, focusing on the temporal dimensions of the unfinished airport project.

Anticipation and suspension

Since plans to build an international airport near Nijgadh emerged in the 1990s, the project has been flanked by great promises in terms of regional and national development. The project, declared 1 of 22 'National Pride Projects' by the National Planning Commission, is envisioned to create thousands of jobs, boost tourism, and make Nepal a major aviation hub in South Asia (19.12.2019 *The Kathmandu Post*). In the Nijgadh Region, which has not been able to benefit from tourism as much as the nearby Chitwan and Parsa National Parks and where more and more people are migrating abroad in search of work, the prospect of employment and economic growth has been of great appeal to many.

People's aspirations have been fuelled by soaring superlatives and stories of future benefits from politicians and project officials, which are circulated in the media and at public meetings. It almost seems as if the longer the delay lasts, the more vigorous the assurances of project proponents are. The Second International Airport was declared the 'most important project in Nepal's history' by its chief engineer and described as 'unavoidable' for national development by former minister of culture, tourism, and civil aviation Yogesh K. Bhattarai (Dhungana 2019; 5 October 2020, *Spotlight Nepal*).

The journalists Rai and Adhikari refer to such repeated assurances in their podcast episode on the SIA as a 'virtuous cycle of infrastructure, tourism, jobs, growth, tourism, jobs, infrastructure, growth' (19.08.2019 *The Record*). This is also reflected in the latest National Five-Year Plan, which claims that the airport will contribute 'to the social and economic transformation of Province No. 2 and the entire country by creating jobs, increasing production and productivity, and facilitating trade and expansion' (National Planning Commission 2020: 509).

Economy of anticipation for the Second International Airport 31

As Braun describes, it is through affective narratives like these that an infrastructure project is 'buoyed' when physical structures are yet to be built (Braun 2020: 866). They give proponents reassurance of promising futures and minimise the space for doubt and concern via the apparent inevitability of the project. Nearly all my respondents from Bara district acknowledged the project's prospective positive impact on the regional and national economies. While some were unsure regarding the extent to which the project would improve their personal lives, they frequently asserted that they were aware of the airport's greater benefits for their fellow citizens.

Such responses reflect the power of promise and anticipation in shaping people's perceptions of the airport and its relevance to their own livelihoods. In the following sub-sections, I describe how the expected arrival of the airport, fuelled by an 'economy of anticipation', has already affected the lives of residents, both affectively and materially. However, I also show how anticipation is dampened through the prolonged delay of the project and how its 'suspended present' is already having noticeable consequences.

Traces of the airport

When I first arrived in the Nijgadh area, I was constantly on the lookout for potential signs of the airport. In addition to a few hand-painted project signs along the Simara-Nijgadh road and the Fast-Track construction site, the previously mentioned Tara Inn Airport Hotel close to Nijgadh's main market was the first clue to the airport's existence that I came across (see Figure 2.3).

The hotel was probably also one of the most obvious signs of economic aspirations that the airport project had incited. Nijgadh has long been a typical Terai highway town, overlooked by domestic and international tourists. Until a few years ago, there was no commercial hotel. As one respondent from Tangiya Basti mentioned, 'if guests came from Kathmandu, [they] had to take them to hotels in Birgunj or Simara or Hetauda' (respondent from Tangiya Basti, interview transcription, 24 February 2020).

Figure 2.3 Visible traces of the SIA project in Bara district: (a) project sign along the East–West Highway and (b) SIA project office in Simara.

Source: Photograph by the author Geschewski (2021).

This has since changed: when I searched for accommodation during my fieldwork, I could choose from at least three hotels. Hopes that the airport and Fast Track would direct more visitors to Nijgadh have already translated into actual private investments, especially since it was decided that there would not be a separate airport city (20.12.2018 *The Kathmandu Post*). These developments have not gone unnoticed. As one respondent from Tangiya Basti remarked: 'Look, in anticipation of the airport, they are building big hotels and setting up businesses' (respondent from Tangiya Basti, interview transcription, 24 February 2020).

While Nijgadh is not yet frequented by tourists, the hotels have become a key venue for another economic activity that the anticipated airport has boosted: real estate speculation. Every evening, when I returned to my hotel after a day of fieldwork, I saw new groups of people arriving from Kathmandu and other parts of the country. As hotel staff and residents confirmed in informal conversations, these people had come to investigate opportunities for land purchases, attracted by the promises of Nijgadh's presumed bright economic future.

Land prices have skyrocketed since plans for the construction of the SIA and the expressway gained momentum. According to several respondents, in some rural areas, the price of a *kattha* (about 340 m^2) has risen from NPR[3] 3 lakh (EUR 2,160) to over NPR 20 lakh (EUR 14,435) in less than a decade, and in the outskirts of Nijgadh, prices now start at NPR 25 lakh per *kattha*. The closer to the airport, the main market of Nijgadh, and the access roads land is, the more prices have soared. Land is not only sold for business purposes, for example, to build more hotels, shops, and restaurants for future tourists. Rather, it has become a speculative commodity for the affluent. This can also be observed on Nepali real estate websites, where private sellers and real estate agents offer plots near the airport, Nijgadh, and the expressway as investment opportunities (good investment, good money), promising that the land's value will drastically increase within a very short time frame (see Figure 2.4).

The particular increase in land prices in the Nijgadh Municipality, even though the airport is located in the neighbouring Kolhabi Municipality, may also be related to the fact that the Second International Airport was commonly referred to as 'Nijgadh International Airport'. As several respondents, including the quoted resident of Tangiya Basti, noted, this name was likely used intentionally:

> Because of the name, land prices in Nijgadh are going up. All the 'big shots'[4] come from Nijgadh, so they named the airport after that place. People come to buy land in Nijgadh because they think the airport will be built there, but that is not the case.
> (respondent from Tangiya Basti, interview transcription, 27 February 2020)

Another respondent pointed to the role of the local media in helping to keep anticipation for the airport high:

> People of Nijgadh want to drive up the price of their land. That's why they spread all this information . . . that an airport is going to be built here. When the noise subsides, it is the media of Nijgadh who bring it up again.

Economy of anticipation for the Second International Airport 33

LAND ON SALE
Good investment good money

-> Near from fast track and upcoming international airport project.

->You can get now on affordable price but after some months and year it'll be hard for you to buy.

Note:- Nice opportunity for investing on land which can get double or even beyond your thinking after some months and year

* Just 10-12 km far from Fast track

* And also International airport is being construed.

Contact on:- 98

Figure 2.4 Online advertisement for real estate in Bara district.
Source: HamroBazaar (2021).

Cross describes the escalation of land prices in the wake of a promised infrastructure project as 'speculative exuberance', an ample demonstration of how anticipation is used to transform local landscapes and economies and how this 'economy of anticipation' produces winners and losers (Cross 2015: 430). In the case study, the beneficiaries are mainly local landowners and brokers, colloquially described as a 'land mafia', which illustrates their economic and political influence.

The less fortunate are those who need land and can no longer afford it, such as the expropriated people of Kathgaun.

Victims of land speculation

Most residents of Kathgaun agreed to the compensation rate set by the government in 2017, which ranged from NPR 2.41 lakh and NPR 4.6 lakh per *kattha*,[5] depending on the distance to the nearest road (29.03.2017 *The Kathmandu Post*). After being told that they had only 35 days to apply for compensation and that, otherwise, they would not receive any support, many people readily accepted. In addition, they did not want to stand in the way of the progress the airport promised to bring, accepting what Braun calls 'local pain for national gain' (Braun 2020: 854). A Kathgaun resident who has handed over his land recalled, 'We thought that if they want to build the airport, which will help the development of the whole nation, why should we stay here and reject this offer?' (respondent from Kathgaun, interview transcription, 4 March 2020).

However, many expropriated residents soon found that the amount of compensation did not match the increased land prices in the region. One resident explained to us, 'Now that everyone knows about the airport, land prices have gone up, so people from here can no longer afford to buy land.' Some former residents have since settled for much smaller land plots than they had previously owned, compromising their ability to sustain themselves through agriculture. Others, unable to find affordable land, have stayed in Kathgaun, reassured by the statements on the part of some local politicians that the construction of the airport will take at least another ten years and that they can remain in their houses until then.

Meanwhile, many of these families have used part of their compensation to cover daily expenses, which has further reduced their chances of acquiring land in the region. One resident explained, 'It's been two years since we took the money. We spent it on everything. We haven't saved any money for the future' (respondent from Kathgaun, interview transcription, 4 March 2020). Pondering the future and where his family would go if the construction of the airport actually began one day, he continued, 'Do you know *banjaras*[6]? We will have to live just like that. If we don't have anything, the only option is to live like them.'

Hopes and demands

While, for many people in Kathgaun, the promises of the airport have already turned to despair, the residents of Tangiya Basti are still negotiating their future. Because the villagers live on government land, they are not entitled to any financial compensation, but they were repeatedly assured by national government officials and project officers that their resettlement would be taken care of. One resident remembered the following:

> Government representatives used to come and tell us that the airport was going to be built here and we would be moved to a different place. They

promised that we were going to get good compensation and were going to be located in a good place.
(resident from Tangiya Basti, interview transcription, 25 February 2020)

However, despite past promises, details such as the exact relocation site, plot size per household, provision of facilities, and assistance in moving and rebuilding houses are yet to be decided (05.09.2019 *My Republica*; 22.07.2020 *The Record*). To counteract the continued uncertainty, the residents have since formulated their own demands for their resettlement, which were disseminated in 2017 by the Tangiya Basti Concern Committee, a local representative body that was founded in 2007 due to the airport issue. Two of the seven demands are (1) resettlement within 3 km of the airport and (b) one guaranteed job per household in the construction and operation of the airport.

Both demands reflect villagers' aspirations about the benefits the airport could create for their lives. The 'economy of anticipation', with promises such as the creation of 100,000 direct and indirect jobs, as stated in the detailed feasibility study of 2011, had also left its mark on Tangiya Basti (19.12.2019 *The Kathmandu Post*). While many respondents agreed that land and business owners in Nijgadh would benefit most from the airport, many hoped that, as long as they could stay close to the airport, they would not be left empty-handed:

Our children could get jobs at the airport. We could sell our vegetables to the hotels that will be built here. If we live nearby, at least we might be able to open small businesses. . . . This is why we demanded for land nearby.
(resident from Tangiya Basti, interview transcription, 27 February 2020)

As expressed in the preceding quote from an older villager, the prospect of stable employment for local youth was compelling. For the younger generation, who are increasingly better educated, working in their families' fields has become less and less attractive. Due to the lack of local employment opportunities, labour migration to urban areas or abroad was often perceived as the only option. A resident in his early 30s who had recently returned from Malaysia shared his view with us: 'Our main source of income is farming or going abroad. There are only two alternatives for us' (resident from Tangiya Basti, interview transcription, 25 February 2020).

Others in Tangiya Basti hoped of obtaining land titles after relocation. Considering the fact that most villagers had been landless all their lives, the prospect of landownership was attractive even to those who felt very attached to the village: 'Once the airport is built, people may finally have a permanent settlement. They will have property deeds. Right now, we don't have any deeds' (resident from Tangiya Basti, interview transcription, 25 February 2020).

The promise of development can be enticing not so much, as Cross states, because it is 'inherently persuasive' but because it 'tap(s) into vernacular dreams for social and material transformation' (Cross 2015: 429). The international airport is attractive to many, including those who face direct losses from it, because it offers a supposed solution to their current problems, not because the project itself is so promising.

Suspension and frustration

However, while some residents of the three settlements remain hopeful, others have always been or have become wary of the project. Some simply did not want to leave their land and lives. As one elderly resident wistfully recounted, 'I think the place will change a lot. Everything will be like in those YouTube videos, but I wish they would just leave it alone and develop this area.' For others, it was not so much because they rejected the airport but because their aspirations were dampened by the continued delay of its execution.

Ever since the first officers came to Tangiya Basti and Kathgaun in 1995/1996 (Matiyani Tol had not yet been established yet) to take soil samples for site assessment, the settlements were frequently visited by politicians and project officials, who always had new plans and promises. Many residents recalled visits by former tourism ministers Prithvi Subba Gurung, Rabindra Adhikari, and Yogesh Bhattarai in 2008, 2018, and 2019, respectively. Each time, it was announced that the airport would be completed within five years. However, since then, not much has happened, as this Tangiya Basti resident soberly noted:

> By now, it feels more like a rumour. I've been hearing about [the SIA] since I was born. A few months ago, a minister [Bhattarai] came and said that they will move us to a new place. They give speeches about completing the airport within two years. We have all stopped believing them. It has been so long.
> (resident from Tangiya Basti, interview transcription, 1 March 2020)

The continued suspension of the airport has not only changed many residents' perceptions of the project but also affected local development, especially in Tangiya Basti and Matiyani Tol. While both villages were previously refused facilities such as electricity and mobile phone networks due to their informal status,[7] the denial has now been justified by the expected beginning of construction. A respondent from Tangiya Basti explained as follows:

> Three or four years ago, we went to Kathmandu and submitted an estimate of 65 lakh [EUR 47,675] to lay a power line here, but the government said no. Why should we supply a line to a place from where you will move? Why should the government waste money?
> (resident from Tangiya Basti, interview transcription, 1 March 2020)

Similar comments about potential 'double expenses' were made by a project official and a local politician I interviewed. While surrounding villages were electrified long ago, people in both settlements use self-financed small solar panels. In January 2020, after years of pleading with the authorities, new hope emerged in Tangiya Basti when the former director of the Nepal Electricity Authority, Kulman Ghising, promised the electrification of the villages (26.01.2020 *The Himalayan Times*). However, as a resident stated, 'the CDO [chief district officer] has stopped the process, saying that there is no reason to invest in a settlement that will be relocated' (resident from Tangiya Basti, interview transcription, 25 February 2020).

In addition to denying the residents basic facilities, the government has also restricted renovation work, as several interviewees stated: 'They tell us not to rebuild, because we will be relocated soon. The head of the airport project told us that we were not allowed to build houses for animal shelters.' Because of the airport project, residents are forced to live in the past, without certainty regarding their future. Similar to other delayed megaprojects, the continued deferral of construction and resettlement decisions has led to 'individual anxiety and limited economic development' (Carse & Kneas 2019: 20), showing the very tangible effect of the 'suspended present' of the SIA as an unfinished infrastructure.

The uncertain future of the airport

As the development of the airport has moved back and forth between anticipation and stagnation, its future is as uncertain as ever. Its completion is blocked by two public interest lawsuits in the Supreme Court, and its financing remains unresolved. While local residents hope for a quick solution for a variety of reasons, the delay is in the interest of airport opponents in Kathmandu. One activist I interviewed explained that, despite the negative impact on residents, the suspension has provided valuable time for protest:

> *For us, it is good. The more we can wait, the more discouraged investors will be, the more attention this project will get. We have more time to bring in the international community. I think the only problem is that the locals [the residents] will be kept in limbo for much longer.*
> (environmental activist from Kathmandu, interview transcription, 11 March 2020)

While airport opponents largely acknowledge the need to expand Nepal's aviation sector, they advocate for the relocation of the airport to an alternative site, such as Murthiya, 35 km farther east, where there would be less environmental damage (27.10.2018 *Spotlight Nepal*). Through the ongoing debates about the SIA and its significance for Nepal's economic development, the inevitability of the project is actively being questioned. This illustrates the potential of suspension to generate 'alternative representations of project completion' (Carse & Kneas 2019: 20).

Alternative paths, however, are in direct contradiction to the ever-higher stakes in the project that the 'economy of anticipation' has brought for some. These stakes do not necessarily depend on the actual operation of the airport but, rather, on the continued possibility of realising the airport in its current form and location. A forestry expert and signatory to one of the public interest litigations I interviewed in Kathmandu recalled an encounter with local businesspeople in Simara, a town around 25 km east of Nijgadh. When he mentioned that his group had come to Bara to inspect alternative sites for the airport, they reacted with concern: 'They were very shocked, the businessmen. They said, "Our investment will be gone. We will suffer a loss"' (environmental scientist from Kathmandu, interview transcription, 17 March 2020).

It becomes clear that the airport project is surrounded by many vested interests that go far beyond the direct benefits and losses of the project. As Rai and Adhikari state, 'The problem now is not even the adhoc-ism, the lack of consultation, serious study, adequate preparation and transparency. It is the enormous political investment in this project' (19.08.2019 *The Record*).

Conclusion

The convoluted genesis of the SIA project shows how an airport project can acquire socio-cultural and material significance long before the foundation is laid. Its physical arrival is preceded by imaginaries that arise from affective narratives of anticipation. Promises of development and modernisation, as well as reservations about the loss of livelihoods and environmental degradation, have influenced people's attitudes towards the project. Anticipatory narratives about the 'soon to come', conceptualised here as the 'economy of anticipation', have transformed local and regional economies, land use and tenure systems, and social and political relationships, already producing winners and losers.

In the case of the SIA, however, the anticipation of what is to come is also obscured by what is not there. Affective notions of hope and longing conflict with the emergence of disillusionment, fear, and resistance. While much attention has been paid to the future positive and destructive impacts of the airport and its role in national development, far less emphasis has been placed on the impact of the long delay on communities that are in a seemingly endless state of limbo. Similar to other unfinished megaprojects, the SIA's suspended present has created space for economic opportunity and speculation for some, but for others, it has jeopardised local development. Nijgadh residents who sold their land at inflated prices and real estate agents have already benefited from the project. Private investors and business owners are still waiting for their turn to profit. For the people in the three affected villages, the bar is much lower because they merely long for security after years of uncertainty.

Overall, approaching the case study through a temporal lens has illustrated the meaning of time and waiting in a significantly delayed infrastructural project cycle. While politicians and project authorities stress the alleged future benefits of the airport, they tend to omit the years of suspension that have put thousands of residents in a prolonged limbo, depriving them of past and present opportunities and rendering their future uncertain.

To date, it is unclear whether the airport will ever materialise. In the absence of a secure financing concept, and with international air traffic undergoing major shifts due to increased climate awareness and changed mobility patterns as a result of the coronavirus pandemic, it remains to be seen how long the airport's proponents can continue to make the case for their project. The proclaimed urgency of the airport may weaken further with the imminent opening of the Pokhara and Bhairawa airports. While it is difficult to assess the consequences of a potential abandonment of Nijgadh, it is certain that it would be hailed by some as a victory for *sustainable* development, while for many in the Nijgadh region, it would destroy the dream of their interpretation of development.

Acknowledgements

I would like to thank the residents of Tangiya Basti, Kathgaun, and Matiyani Tol for sharing their stories and knowledge. I also thank Dr Dil Bahadur Khatri, Kaustuv Neupane, Dr Sabin Ninglekhu, and Dr Andreas Johansson from the Swedish South Asian Studies Network for their input and support.

Notes

1 The Nepali Civil War, also referred to as the Maoist Insurgency, began in 1996 when Maoist groups, under Pushpa Kamal Dahal, launched a 'people's war' to overthrow the monarchy and turn Nepal into a one-party communist republic. After ten years and over 13,000 deaths, the Maoists and the newly sworn-in prime minister Girija Prasad Koirala signed a peace agreement in November 2006 that ended the war.
2 *Taungya*, derived from the Burmese terms *taung* for 'hill' and *ya* for 'cultivation', describes a type of shifting cultivation in agroforestry. The Taungya system was introduced in Nepal in 1972–1974 in the Tamagadhi area near Nijgadh, indicating that the founders of Tangiya Basti were among the first in Nepal to use it (see Geschewski & Islar 2022) for more information).
3 NPR = Nepali rupee.
4 Referring to local influential politicians.
5 Between EUR 5 and EUR 10 per m^2.
6 Hindi term for 'nomad'.
7 Legally, informal settlements are not entitled access to basic facilities, such as water supply, sanitation, electricity, and infrastructure, and are often excluded from government development initiatives (Shrestha 2014).

References and sources

Adams, V., M. Murphy, and A. Clarke (2009). 'Anticipation: Technoscience, life, affect, temporality'. *Subjectivity* 28, 246–265. https://doi.org/10.1057/sub.2009.18

Appel, H., N. Anand, and A. Gupta (2018). 'Introduction: Temporality, politics, and the promise of infrastructure'. In *The promise of infrastructure*. H. Appel, N. Anand, and A. Gupta (eds.). New York: Duke University Press, 1–38. https://doi.org/10.1515/9781478002031-002

Bhattarai, K. and D. Conway (2021). *Contemporary environmental problems in Nepal: Geographic perspectives*. Cham: Springer. https://doi.org/10.1007/978-3-030-50168-6

Braun, Y. (2020). 'Lesotho's white gold: The political ecology of temporality and the economy of anticipation in resource extraction and large dam infrastructural projects'. *Journal of Political Ecology* 27(1), 853–876. https://doi.org/10.2458/v27i1.23250

Bridger, R. (2017). Nijgadh aerotropolis – 2.4 million trees could be felled, and 7,380 people displaced. 4 October. Retrieved 15.06.2021. https://antiaero.org/2017/10/04/nijgadh-aerotropolis-2-4-million-trees-could-be-felled-and-7380-people-displaced/

CAAN (2020). *Civil Aviation annual report 2019–20*. Kathmandu: Civil Aviation Authority of Nepal (CAAN). Retrieved 10.05.2021. https://caanepal.gov.np/storage/app/uploads/public/5ff/2c3/8de/5ff2c38de0957043775720.pdf

Carse, A. and D. Kneas (2019). 'Unbuilt and unfinished: The temporalities of infrastructure'. *Environment and Society* 10(1), 9–28. https://doi.org/10.3167/ares.2019.100102

Chernaik, M. (2019). *On biological resources permanently lost to the Second International Airport project in Kolhabi municipality in Bara district of Province 2*. Eugene, OR: Environmental Law Alliance Worldwide U.S.

Cross, J. (2015). 'The economy of anticipation: Hope, infrastructure, and economic zones in South India'. *Comparative Studies of South Asia, Africa and the Middle East* 35(3), 424–437. https://doi.org/10.1215/1089201X-3426277

Delgado, A. L. (2018). *Sumaq Kawsay, Allin Kawsay: Conceptions of well-being among quechua female vendors in the face of change in Chinchero, Peru*. MA thesis. Vanderbilt University, Nashville. https://ir.vanderbilt.edu/handle/1803/10657?show=full

Dhungana (2019). Nijgadh International – The Record docs. Retrieved 12.12.2020. www.youtube.com/watch?v=pXNpwVn6lQw

Fent, A. and E. Kojola (2020). 'Political ecologies of time and temporality in resource extraction'. *Journal of Political Ecology* 27(1), 819–829. https://doi.org/10.2458/v27i1.23252

GEOCE (2018). *Environmental impact assessment (EIA) of Second International Airport project*. Kathmandu: GEOCE Consultants (P) Ltd.

Geschewski, H. (2021). *"They only talk about the trees" – An analysis of relations of power and justice in the (de)construction of Nepal's Second International Airport*. Master's thesis. Lund University.

Geschewski, H. and M. Islar (2022). 'A political ecology of aviation and development: An analysis of relations of power and justice in the (de)construction of Nepal's Second International Airport'. *Journal of Political Ecology* 29(1), 51–75. https://doi.org/10.2458/jpe.2304

Haines, S. (2018). 'Imagining the highway: Anticipating infrastructural and environmental change in Belize'. *Ethnos* 83(2), 392–413. https://doi.org/10.1080/00141844.2017.1282974

HamroBazaar. (2021). Land on Sale near Nijgadh Airport. Retrieved 21.07.2021.

Harris, T. (2021). 'Air pressure: Temporal hierarchies in Nepali Aviation'. *Cultural Anthropology* 36(1), 83–109. https://doi.org/10.14506/ca36.1.04

Hetherington, K. (2014). 'Waiting for the surveyor: Development promises and the temporality of infrastructure'. *The Journal of Latin American and Caribbean Anthropology* 19(2), 195–211. https://doi.org/10.1111/jlca.12100

Investment Board Nepal (2019). *Nijgadh International Airport*. Kathmandu: Office of the Investment Board Nepal. Retrieved 04.08.2020. https://investmentsummitnepal.com/wp-content/uploads/2019/03/Nijghad-International-Airport-Bara-1.pdf

Khatiwada, S. and R. Aryal (2019). 'Effective implementation of the "Kathmandu – Terai Fast Track" road project'. In *KEC Conference 2019*, Lalitpur, p. 8.

Kirchherr, J., T. Pomun, and M. Walton (2016). 'Mapping the social impacts of "Damocles Projects": The case of Thailand's (as yet Unbuilt) Kaeng Suea Ten Dam'. *Journal of International Development* 30(3), 474–492. https://doi.org/10.1002/jid.3246

National Planning Commission (1980). *Sixth plan (1980–1985)*. Kathmandu: Government of Nepal. Retrieved 05.08.2020. www.npc.gov.np/images/category/sixth_eng.pdf

National Planning Commission (2002). *Tenth plan (2002–2007)*. Kathmandu: Government of Nepal. Retrieved 30.07.2020. www.npc.gov.np/images/category/10th_eng.pdf

National Planning Commission (2020). *15th plan (FY 2076/77–2080/81)*. Kathmandu: Government of Nepal. Retrieved 10.08.2020. www.npc.gov.np/images/category/15th_Plan_Final1.pdf

NEFEJ (2019). *Report on environmental, social and economic impacts of Second International Nijgadh Airport*. Lalitpur: Nepal Forum of Environmental Journalists. Retrieved 03.08.2020. https://nefej.org.np/1043/

Okebugwu, O. and E. Omajeh (2014). 'Debunking the 'White Elephant Project' myth'. *Project Management World Journal* 3(5), 1–9.

Rest, M. (2019). 'Dreaming of pipes: Kathmandu's long-delayed Melamchi water supply project'. *Environment and Planning C: Politics and Space* 37(7), 1198–1216. https://doi.org/10.1177/2399654418794015

Schindler, S., S. Fadaee, and D. Brockington (2019). 'Contemporary megaprojects: An introduction'. *Environment and Society* 10(1), 1–8. https://doi.org/10.3167/ares.2019.100101

Shah, S. G. (2019). 'An analysis of EIA report of the Second International Airport project, Nepal'. *Hydro Nepal: Journal of Water, Energy and Environment* 24, 57–67. https://doi.org/10.3126/hn.v24i0.23585

Shakya, P. (2019). 'Kathmandu-Terai/Madhesh Fast Track Road project, Nepal'. *Environmental Justice Atlas*, 18 December. Retrieved 23.08.2020. https://ejatlas.org/conflict/kathmandu-terai-madhesh-fast-track-road-project-nepal

Shrestha, R., J. Zevenbergen, M. Banskota, and A. Tuladhar (2014). '"Decades of struggle for space": About the legitimacy of informal settlements in urban areas'. In *FIG Conference*, Kuala Lumpur. Retrieved 15.08.2020. www.fig.net/resources/proceedings/fig_proceedings/fig2014/papers/ts01g/TS01G_tuladhar_zevenbergen_et_al_7095.pdf

Thakur, R. K. (2015). *Geospatial modeling of walk trails of Asian Elephant Elephas maximus as tool for mitigating human elephant conflicts in Central Nepal*. The Rufford Foundation. Retrieved 21.08.2020. www.rufford.org/files/17279-1%20Final%20Report.pdf

Media sources

29.03.2017. 'Compensation rate for land for Nijgadh airport fixed'. *The Kathmandu Post*. Retrieved 24.10.2020. https://kathmandupost.com/money/2017/03/29/compensation-rate-for-land-for-nijgadh-airport-fixed

01.06.2018. 'Does Nepal need a 4th international airport?'. *Nepali Times*. Retrieved 10.11.2021. www.nepalitimes.com/banner/does-nepal-need-a-4th-international-airport/

27.10.2018. '"Murtiya will be best alternative to Nijgadh for the construction of Second International Airport": Chanda Rana'. *Spotlight Nepal*. Retrieved 27.10.2021. www.spotlightnepal.com/2018/10/27/murtiya-will-be-best-alternative-nijgadh-construction-second-international-airport-chanda-rana/

20.12.2018. 'Airport city not necessary in Nijgadh: House panel'. *The Kathmandu Post*. Retrieved 24.10.2021. https://kathmandupost.com/national/2018/12/19/airport-city-not-necessary-in-nijgadh-house-panel

28.03.2019. 'What does the environmental impact assessment report say about the feasibility of the Nijgadh Airport project?'. *The Record*. Retrieved 25.08.2020. www.recordnepal.com/category-explainers/what-does-the-environmental-impact-assessment-report-say-about-the-feasibility-of-the-nijgadh-airport-project/

16.07.2019. 'Nijgadh Airport: What is proposed and why environmental analysts are so worried'. *The Record*. Retrieved 04.01.2021. www.recordnepal.com/category-explainers/nijgadh-airport-what-is-proposed-and-why-environmental-analysts-are-so-worried/

18.07.2019. 'Life in limbo for the residents of Tangia Basti'. *The Record*. Retrieved 03.08.2020. www.recordnepal.com/wire/photo-essays/life-in-limbo-for-the-residents-of-tangia-basti/

19.08.2019. 'Oversold, overhyped, evidence-free development'. *The Record*. Retrieved 06.09.2020. www.recordnepal.com/podcast/oversold-overhyped-evidence-free-development/

05.09.2019. 'Rs 600 million distributed to project-affected so far'. *My Republica*. Retrieved 24.10.2020. http://myrepublica.nagariknetwork.com/news/73672/

24.09.2019. 'Zürich Airport shortlisted to develop Nijgadh Airport'. *The Himalayan Times*. Retrieved 17.08.2020. https://thehimalayantimes.com/business/zurich-airport-shortlisted-to-develop-nijgadh-airport/

19.12.2019. 'Nepal needs another international airport. So why is Nijgadh being held up?'. *The Kathmandu Post*. Retrieved 03.08.2020. https://kathmandupost.com/province-no-2/2019/12/19/nepal-needs-another-international-airport-so-why-is-nijgadh-being-held-up

26.01.2020. 'Directive to electrify Tangiyabasti'. *The Himalayan Times*. Retrieved 02.10.2020. https://thehimalayantimes.com/nepal/directive-to-electrify-tangiyabasti/

13.02.2020. 'Nijgadh: Prospects and priorities'. *The Record*. Retrieved 13.08.2020. www.recordnepal.com/perspective/nijgadh-prospects-and-priorities/

22.07.2020. 'For the airport yet to come'. *The Record*. Retrieved 26.08.2020. www.recordnepal.com/wire/features/for-the-airport-yet-to-come/

17.08.2020. 'Nijgadh International Airport debate: These are key arguments of both sides'. *OnlineKhabar*. Retrieved 10.11.2021. https://english.onlinekhabar.com/nijgadh-international-airport-debate-these-are-key-arguments-of-both-sides.html

23.09.2020. 'Flughafen Zürich: Kein Interesse mehr an Nepal-Projekt'. *Handelszeitung*. Retrieved 30.12.2020. www.handelszeitung.ch/panorama/der-flughafen-zurich-verliert-das-interesse-nepal

05.10.2020. 'Nijgadh airport row: Petitioners to file a contempt of court against tourism minister Bhattarai". *Spotlight Nepal*. Retrieved 19.01.2021. www.spotlightnepal.com/2020/10/05/nijgadh-airport-row-petitioners-file-contempt-court-against-tourism-minister-bhattarai/

06.10.2020. 'Multi-billion-dollar Nijgadh Airport project hangs in the balance'. *The Kathmandu Post*. Retrieved 04.01.2021. https://kathmandupost.com/national/2020/10/06/multi-billion-dollar-nijgadh-airport-project-hangs-in-the-balance

03.02.2021. 'Government backs plan to build Nijgadh airport'. *The Kathmandu Post*. Retrieved 04.10.2021. https://kathmandupost.com/money/2021/02/03/government-backs-plan-to-build-nijgadh-airport

29.05.2021. 'Nepal budget 2021/22: Addressed the few aviation programs'. *Aviation Nepal*. Retrieved 09.07.2021. www.aviationnepal.com/nepal-budget-2021-22-addressed-the-few-aviation-programs/

30.05.2021. 'Government aims to operate new international airports in the next fiscal year'. *My Republica*. Retrieved 04.10.2021. http://myrepublica.nagariknetwork.com/news/111587/

27.05.2022. 'Supreme Court says no to Nijgadh airport over environmental concerns'. *The Kathmandu Post*. Retrieved 15.01.2024. https://kathmandupost.com/money/2022/05/27/supreme-court-says-no-to-nijgadh-airport-over-environmental-concerns

12.06.2023. 'Despite criticism, Nijgadh International Airport once again gets 'symbolic funds' in new budget'. *OnlineKhabar*. Retrieved 15.01.2024. https://english.onlinekhabar.com/nijgadh-international-airport-budget.html

09.11.2023. 'Govt initiates preparation for construction of Nijgadh airport'. *My Republica*. Retrieved 15.01.2024. https://myrepublica.nagariknetwork.com/news/govt-initiates-preparation-for-construction-of-nijgadh-airport/

10.11.2023. 'Govt decides to move forward with Nijgadh International Airport construction'. *OnlineKhabar*. Retrieved 15.01.2024. https://english.onlinekhabar.com/nijgadh-international-airport-goahead.html

3 The rise of infrastructure-induced human–elephant conflict in Sri Lanka

A case study of Mattala Rajapaksa International Airport

Menusha Gunasekara and Dishani Senaratne

Introduction

If you are looking for a vacation destination with sun and mesmerising wildlife, Sri Lanka – the pearl of the Indian Ocean – has a lot to offer. As wild elephants are one of the great attractions in the country, watching wild elephants at the national parks in Hambantota[1] should be necessary on your bucket list. Boosting eco-tourism is one of the reasons for backing the establishment of the Mattala Rajapaksa International Airport (MRIA, popularly known as Mattala Airport) as the second international airport in Sri Lanka. However, the airport has been making headlines not for its success but for the environmental conflicts it has caused.

Prior to Mattala Airport construction, Hambantota had experienced human–elephant conflict (HEC) to some degree since the construction of the Walawe left-bank settlements. This will be discussed in detail throughout the chapter. During the construction period of the airport, wildlife and forest destruction was severe, yet the local farming community did not perceive it as a 'wake-up call' regarding potential environmental concerns that could arise in the near future. Now, after almost a decade, a new set of problems has manifested because of the airport. The intensification of HEC in Hambantota, which has risen at an alarming rate, is mobilising the local community to demand solutions. Complicating the situation, the land value around the airport has now skyrocketed, attracting illegal encroachment. State-led resource exploitation is further driving this land grabbing. By unravelling and examining the relationships between the land, airport, wildlife, and politics, this chapter aims to shed light on contestations at the airport site.

This chapter shows that political ecology provides a useful lens for examining the relationships between environmental destruction, social marginalisation, and access to and control over resources, which have produced environmental struggles (Bryant & Bailey 1997; Delang 2005; Camisani 2018). A common premise in political ecology is that environmental changes are the product of political process (Bryant and Bailey 1997; Robbins 2012). This is based on three fundamentally linked assumptions outlined by Bryant and Bailey (1997) and which guided this research. The first assumption is that the costs and benefits of environmental change are asymmetrically distributed among the involved groups. The second is that the unequal distribution of the costs and benefits of ecological change either

DOI: 10.4324/9781003494966-3

reinforce or reduce existing socioeconomic inequalities. The third is that the asymmetric distribution of costs and benefits and their effects on existing inequalities have political implications that can modify power relationships (Bryant & Bailey 1997: 27–28). With these assumptions in mind, this study aims to uncover sociopolitical and ecological implications from the case of Mattala Airport.

Located on the southeastern coast of Sri Lanka, the Hambantota district belongs to the Southern Province. Hambantota is the main town of the district. With a total land area of 2,609 km^2, Hambantota has traditionally been a Sinhalese-majority district with a population of 599,903. While the district is primarily an agricultural region, the unemployment rate (7.3%) is higher than the national rate of 4% (DCS 2012, 2019). Lying within the Hambantota and Moneragala districts, Yala National Park is the country's second-largest national park, drawing locals and tourists alike. The beach at Rekawa is another popular tourist destination, known to attract several endangered species of sea turtles to lay their eggs.

The empirical section of the chapter, which follows the methodology, is divided into two parts. First, we outline the 2013 establishment of Mattala Airport during the Mahinda Rajapaksa regime, which envisaged transforming Hambantota into a new urban hub; a discussion of the status of the airport follows (Section 3). Next, we explore the environmental impact of the airport, with a special focus on the Walsapugala farmers' protest in juxtaposition with the unequal distribution of the costs of airport construction and the alleged land grabbing (Sections 4–6). We argue that the communities are no longer demanding cash handouts; they are calling for government action in elephant conservation rather than being deluded by the construction of large-scale infrastructure in traditionally agrarian areas.

Methodology

This case study employs both primary and secondary data. Initial contact with participants was obtained through an independent regional journalist, who assisted in contacting other participants. Telephone-based semi-structured interviews with six participants were conducted in 2021, respecting travel restrictions due to the Covid-19 pandemic. Our interview partners included officers of agrarian organisations, female farmers, and a male youth, who all participated in the Walsapugala farmers' protest. Interviews were conducted and recorded in Sinhala and later transcribed into English. Interview transcriptions were coded, and themes were identified (Bryman 2012). Secondary data was gathered from a wide range of published and unpublished scholarly and grey literature, including reports by the government, non-governmental agencies, and wildlife research organisations, as well as news reports.

The establishment of Mattala Airport

Constructed in 1934, the Colombo Ratmalana International Airport was Sri Lanka's first international airport, until the inauguration of the Bandaranaike International Airport (BIA) in Katunayake. Initially called Katunayake International Airport and later renamed after former prime minister S. W. R. D. Bandaranaike (1952–1956), the BIA began as a base for the United Kingdom's Royal Air Force during the Second World War. Presumably with minimal attention paid to the environmental impact and the voice of local communities, the land acquisition of both airports would have been expedited under British rule. Located within close proximity to the commercial hub of Colombo, both airports were situated in relatively urban areas at the time of their construction.

The emergence of political discourse on the need for an alternative international airport to the BIA dates back to the 1970s. From 1970 to 2006, nine sites in various areas of the country were considered for this purpose (Attanayake 2018). The Chandrika Bandaranaike Kumaratunga administration (1994–2005) spearheaded the Hambantota City Development Project (1994) that proposed to build a seaport and an airport. This project was driven by the idea that the implementation of a large-scale development project could help alleviate poverty in Hambantota and the neighbouring districts (AGD 2015). However, this ambitious plan never materialised, partly because of governmental changes against the larger backdrop of escalating violence between the national security forces and the Liberation Tigers of Tamil Eelam (LTTE, also called the Tamil Tigers).[2]

With Mahinda Rajapaksa's ascent to presidency in 2005, political conversations about a new international airport were revived. Taking a step further, the Greater Hambantota Development Plan was proposed in 2007 in the aftermath of the devastating 2004 tsunami and subsequently launched in 2011 (AGD 2015). The plan envisaged transforming Hambantota, which was lagging behind in infrastructure development, into a hub through the establishment of a seaport, an airport, and an administrative complex, among other amenities.[3] Initially, the town of Weerawila, where there was already a military airport, was earmarked to be developed as the site of the new airport, despite being located within close proximity to Bundala National Park, the most important birding destination in Sri Lanka. In response to community-led protests, the government shifted the construction of the airport to Mattala. Paradoxically, the proposed site in Mattala is located in close proximity to another four wildlife and forest habitats: Kumana, which hosts a bird sanctuary; Udawalawa, which provides an important habitat for waterbirds and elephants; Lunugamwehera, which is another important habitat for waterbirds and elephants; and Andarawewa Usgala, which is a forest reserve and a crucial corridor for elephants. Attanayake (2018) pointed out that the environmental impact assessment (EIA) of Mattala Airport failed to identify potential issues that could arise from wild animal movement while focusing exclusively on prospective boons to tourism.

With Chinese monetary loans, technical expertise, and labour, the construction of Mattala Airport started in 2009.[4] The airport was officially opened in 2013, touted as having a long runway able to handle the largest commercial jets, among other innovative features. In keeping with the Mahinda Rajapaksa administration's focus on infrastructural development in Sinhalese-majority areas in the post-conflict afterlife, the international airport was projected to be an employment generator, especially for local youth. In contrast, the implementation of mega-development projects like 'Resurrection of the North' (*Uthuru Wasanthaya)* and 'Reawakening of the East' (*Nagenahira Navodaya*) downplayed the need to address the desires of the Tamil community, especially those living in former war zones, to bring about long-term peace.

Apparently swayed by such political rhetoric, the local communities raised minimal objection to the construction of the airport. On the one hand, numerous billboards acknowledging President Mahinda Rajapaksa – largely put up by his prominent political supporters for his taking the initiative to develop Hambantota, the ancestral home of the Rajapaksa family – became a ubiquitous sight in the area before and after the establishment of the airport and other new infrastructure projects (see Figure 3.1). Previously, following the 2009 military defeat of the LTTE, billboards elevating President Mahinda Rajapaksa to a modern-day monarch had been put up by his political allies to celebrate his taking the leadership to defeat the LTTE. These posters presented him as akin to the legendary Sinhalese king Dutugemunu, who hailed from southern Sri Lanka and defeated the South Indian king Elara and became a common sight across Sinhalese-majority areas, including the Hambantota district. On the other hand, speculation was rife that the international airport in Katunayake would eventually shut down, paving the way for the advent of Mattala Airport as the exclusive international airport in the country.

While billboards indicated the political nature of the airport construction, the naming of Sri Lankan airports has also been subject to politics. In addition to

Figure 3.1 A man passes by a billboard put up by the Airport and Aviation Services Limited featuring the president Mahinda Rajapaksa for accelerating the development of Mattala Airport.

Source: Associated Newspapers of Ceylon Limited.

naming the seaport in Hambantota after the then president Mahinda Rajapaksa, the name refers to the ancient Sinhalese kingdom of Magam Ruhunupura in the south, leveraging dominant Sinhalese Buddhist historical narratives to further political propaganda.[5] Similarly, Mattala Airport is named after Mahinda Rajapaksa, alongside an allusion to the locality, while the first two letters of the acronym of the airport (MR) are an implicit reference to the initials of Mahinda Rajapaksa.[6]

Contrary to the expectations of the Mahinda Rajapaksa administration, Mattala Airport ultimately failed to attract international airlines and passenger traffic. Mattala Airport was subsequently dubbed 'the emptiest airport in the world' by *Forbes Magazine* in 2016 (Forbes 2016). The rare arrival of flights was often featured on state-owned media, especially TV stations, that became the subject of social media users' ridicule. In such scenarios, Mattala Airport was perceived by detractors as a symbol of China's debt trap in Sri Lanka. In 2014, the E01 Expressway (commonly known as the Southern Expressway) was extended to Matara, and later to Hambantota in 2020, in a bid to link the airport and the seaport. Despite such facilitation of road networks, the airport failed to become an aviation hub. Instead, Mattala Airport made headlines for bird strikes and the presence of elephants in the locality (for environmental impacts, see Section 4). Whether surprisingly or not, since its inception, Mattala Airport has become a popular stopover destination for local tourists en route to visiting southern Sri Lanka to see the huge airport buildings; subsequently, the airport has become a tourist destination in its own right.

In 2015, the coalition government led by Maithripala Sirisena, leader of the Sri Lanka Freedom Party (SLFP), and Ranil Wickremesinghe, leader of the United National Party (UNP), defeated the seemingly invincible Mahinda Rajapaksa in the presidential election, pledging to usher in good governance (*yahapalanaya*). Soon after assuming power, the newly appointed unity government decided to cease the operation of SriLankan Airlines at Mattala, citing the national carrier's debt crisis.[7] The image of Mattala Airport was further damaged by the government's decision to store paddy at its unused stores complex. This was a textbook case of how megaprojects are left in abeyance as pet projects of the political predecessor rather than considering them national assets that could contribute to economic development.

Sri Lanka's aviation sector is also illustrative of how China and India vie for presence and influence in the country. In a surprising move, a potential joint venture with India to operate Mattala Airport – which was built with Chinese loans – was proposed by the unity government in 2018. As the power struggle between India and China increased in the country, the Gotabaya Rajapaksa government suspended the joint venture in 2020. In 2019, Palai Airport, located in Jaffna, was renamed Jaffna Airport, the country's third international airport, which was developed with Indian assistance. Placing the Sinhala name of the airport below the Tamil name on the signboard created an uproar on social media networks that brought into focus the wider Sinhalese society's intolerance of linguistic plurality.

While the airport's fate seemed to be that of a white elephant, an abandoned mega-development project, the Covid-19 pandemic finally increased passenger numbers. With the rise in repatriation flights in 2020, Mattala Airport 'woke up' from its dormant status. Offering a ray of hope for government allies, Mattala Airport was poised to be transformed into a commercially busy and viable facility in South Asia through the implementation of an integrated five-year plan. In addition, political conversations on developing a new domestic airport at Batticaloa emerged. Apart from these, the BIA in Katunayake is poised for a major expansion with the proposed construction of a new terminal to provide improved passenger experience and more efficient cargo handling. Thus, the aviation sector in Sri Lanka is growing, which entails consequences for the environment and populations in the areas surrounding the airports.

Environmental damages and their long-term implications

There has been a wide range of ecological destruction on Mattala Airport land, with 8 km^2 of forest turned into airport buildings and another 12 km^2 used for environmental facilities and other infrastructure (AGD 2015). Establishing the airport in a wildlife habitat has become the cause of wildlife conflicts, such as HEC. This is mainly due to the failure of the EIA[8] to identify potential negative environmental impacts of the airport project. Research has shown that the EIA of Mattala Airport clearly lacked scientific analysis of wildlife behaviour, especially regarding the ranging habits of elephants. Likewise, an impact analysis on water sources and airport noise, as well as a cost–benefit analysis of ecological value versus economic benefits, was not conducted. In turn, the gap between an EIA report's stated potential environmental implications and the reality on the airport land is enormous. Thus, it has been argued that the report was 'sugar-coated' to make Hambantota seem suitable for the smooth implementation of the airport project (Peiris 2019).

During the establishment of the airport, construction companies ignored the EIA's wildlife management plan. To clear the forest cover for buildings, large areas were set on fire instead of trees being removed using machinery. Elephants fled from the fires to the villages. Electric fences were erected to trap wild animals on the airport land. This was followed by a mass slaughtering for meat production. Yet the Central Environmental Authority, who administered the EIA poorly, engaged in sanctioning these malpractices. The electric fences around the airport land further obstructed elephants' movements. The planned elephant corridor was reduced from 500 m to 125 m by the Aviation Authority of Sri Lanka (Fernando et al. 2015; Sunday Times Plus 2010).

In the post-construction period, wildlife conflicts, such as bird strikes and the presence of elephants, have been frequent (Sunday Times 2014; Peiris 2019). This is because the airport land remained a breeding and resting ground for local and migratory birds in close proximity to the bird sanctuary of Kumana. The water sources attracted elephants (Peiris 2019). According to media, authorities tried to chase wildlife by depriving them of water and food, as well as employing troops to chase away animals (Columbo Gazette 2016).

A salient but long-term implications of Mattala Airport construction has been the escalation of HEC in Hambantota (AGD 2015; Peiris 2019). Such conflict imposes a significant cost on properties and threat to human safety, as well as threaten the existence of wild elephants too (Fernando et al. 2005, 2011, 2020). Approximately 400 elephants, who suffered from habitat loss and habitat fragmentation due to deforestation, the construction of the airport and the road to Lunugamwehera, as well as due to human encroachment to protected areas, now roam in the district (Fernando et al. 2011, 2005, 2020; Prakash et al. 2020). Mattala Airport construction also set other dynamics in motion that led to land grabbing, such as large-scale banana cultivation and solar power projects (Daily Mirror 2020). The relevance and urgency of the HEC are illustrated by the fact that, in 2019, Sri Lanka reported the world's highest annual elephant death rate, exceeding 400, as well as the second-highest human death rate (121) from human–elephant conflicts (Prakash et al. 2020). In this light, the Managed Elephant Reserve (MER)[9] has been recognised as a crucial conflict mitigation and elephant conservation strategy for Sri Lanka (DWC 2006; Fernando et al. 2011; Fernando 2015). Yet the elephant reserve in Hambantota has received poor commitment from governments since it was recommended and politically decided upon (Peiris 2019). In 2006, the Department of Wildlife Conservation (DWC) incorporated the elephant reserve in its National Policy for Elephant Conservation and Management, which it implemented incompletely (Fernando et al. 2011). In this policy context, the establishment of the elephant reserve in Hambantota was considered a landmark, as it would provide resources for the animals and safety for the animal and human populations. The reserve would also ease the migration of elephants by connecting the national parks in Bundala, Lunugamwehera, and Udawalawa (Fernando et al. 2015).

Protesting against the long-term effects of Mattala Airport

We turn now to the second empirical section of the chapter, which examines the motivations of farmers in Hambantota in requesting that the government gazette the elephant reserve. Gazetting is the official publicising of laws or legal content by the state to the public. Only then is the declared content considered to be under the purview of law and policy regulation. It is important to understand that the eruption of farmers' resistance is a direct response to the long-term environmental effects of airport construction on their everyday lives, which remain severe a decade after initial construction. The section begins with an overview of the Walsapugala farmers' protest and then examines the unequal distribution of the costs from aviation-driven human–elephant conflicts and land grabbing, as well as attempts to bring about environmental justice and alter power relations.

The Walsapugala farmers' call for the MER

On 18 January 2021, a local, predominately Sinhalese community of farmers from Walsapugala Village, who were greatly impacted by human–elephant conflicts, began a peaceful protest (*satyagraha*; see later text), demanding that the government gazette the proposed MER in Hambantota (Newsfirst 2021a; EconomyNext 2021). Aside from gazetting, they communicated the following five demands: (1) the installation of a new electric fence around the reserve, (2) driving elephants to the reserve who are trapped inside villages and forest patches, (3) the provision of separate grasslands for cattle farmers, (4) the release of water to the left bank of the Walawe River, and (5) sanctioning and ending illegal land grabbing in Hambantota (Daily Mirror 2020).

Without a concrete response from the government, 86 agrarian organisations, their members and families, organised a large protest that lasted 105 days (Colombo Gazette 2021). This protest also spurred further protests elsewhere in the country by farmers' organisations to support the cause (Newsfirst 2021b). These organisations mobilised more than 2,000 people. Farmers engaged in peaceful sit-ins. Several fast-unto-death[10] campaigns were implemented, in which two women joined on such to mark the International Women's Day. Protesters blocked roads, and vehicle parades conveyed the protesters' demands across the country and internationally (Newsfirst 2021d).

Protesters used unconventional approaches for their non-violent *satyagraha*. For example, they performed rituals and publicly celebrated national days to impart the messages of the protest. On Independence Day in 2021, protesters raised black flags paired with the national flags to depict the darkness in their lives with HEC (The Hindu 2021). Similarly, on Sinhala Tamil New Year's Day, instead of the traditional milk rice symbolising prosperity, they publicly drank salt porridge to demonstrate their resistance and their plight (Lankanewsweb 2021).

Initially, the government tried to thwart these protests by making promises. On 23 January, the then irrigation minister Chamal Rajapaksa promised to meet protesters' demands, and the fast-unto-death campaign was temporally halted. When farmers realised that they had been deceived, the protests resumed (Newsfirst 2021c). Meanwhile, a counter-protest also emerged, possibly as a distraction, by cultivators on the planned elephant reserve. Interviews revealed that the Sooriyaweva police station negotiated the situation and assisted farmers to dilute the voice of the counter-protest.

While the protest was presented as a sudden eruption by the media, interview partners revealed the gradual evolution of the activities. They recalled that, in 2012, the district development committee in Hambantota approved the reserve, but no implementation followed. In 2017, about 15 farmers' organisations protested, yet their plea regarding the elephant reserve went unheard. When the intensity of HEC became critical in 2019, multiple farmers' organisations formed the Walawa Left Bank Joint Farmers' Organisation to seek solutions collectively. During 2019–2020, prior to both the presidential election and the parliamentary general election, politicians in power assured the gazetting of the reserve, but promises were once again forgotten. A leader from a Walsapugala farmers' organisation commented the following in an interview:

> We waited 13 years without a proper answer. As human-elephant conflicts got worse every year. . . . During elections we got promises. But this time we received more support from farmers and communities here to organise the large-scale protest to remind the government that we are still waiting for the answer.[11]

As a result, in January 2021, farmers mobilised a large-scale protest requesting responses from both the political leadership and government authorities. Finally, in April 2021, perhaps with the strong southern voters' base in Hambantota in mind, the present government finally released the gazette, which declared about 240 km^2 of land as the Hambantota MER (DGP 2021).

Yet uncertainties persisted among farmers about the progress of regulating the reserve. Some farmers claimed that the map of the reserve in the gazette contained an outdated map prior to the construction of the previously mentioned Southern Expressway. They felt that the state actors continued to deceive them deliberately without declaring the MER. Farmers realised the need of a legal framework that could assist them in demanding solutions for HEC, as well as preventing land encroachment. According to the leader of the farmers, a committee comprised of district secretaries and farmers' representatives was formed to discuss the remaining activities, such as establishing the electric fencing around the MER and removing illegal occupants. However, no step was taken until December 2021.

The community-led protests drew responses from various quarters. Seventy days into the *satyagraha*, the minister of irrigation and state minister of national security, disaster management, and home affairs, Chamal Rajapaksa, claimed that the people in Walsapugala were funded by non-governmental organisations (NGOs) and 'pretended to fast but slept soundly', adding how he had recently observed a 'marvellous fight between two elephants' and 'had such fun [witnessing this] only because elephants come to the villages' (LankaTruth 2021). Branding the ethnic and religious minorities as dependents on foreign-funded NGOs who would betray the country has long been part of populist rhetoric in political discourses. The ministerial statement carries a bitter irony, as the Sinhalese-led farmers' protests took place within the electorate of the Rajapaksa family. In 2019, most votes were garnered in their favour during the presidential election. The minister's apathetic statements also downplayed the severity of farmers' problems. The *satyagraha* was later called off by the farmers' organisations after a meeting with Namal Rajapaksa, then minister of sports and the son of the prime minister. He, as the apparent successor of the Rajapaksa family, was depicted as a problem-solver of the Gotabaya Rajapaksa administration.

In 2021, it was striking how a wave of protests by various groups erupted in Sri Lanka that centred on diverse issues communicating disagreement with political decisions.[12] The Indo-Sri Lanka People's Cooperation Movement and other civic groups staged demonstrations to show their solidarity with the struggle of farmers in India. During February and March, social media networks were replete with narratives of deforestation and environmental destruction. These reports may have served as a catalyst to strengthen the Walsapugala farmers' protests. Farmers in Walsapugala realised, somewhat belatedly, that sustainable development was key to minimising the risk of environmental damage. They changed their rhetoric and requested an elephant-centric approach in managing the airport land and the solving of existing conflicts between the animals and local communities.

Unequal distribution of cost

Human–elephant conflicts have disproportionally affected small-scale farmers, who have struggled, as their crops and properties have been damaged by elephants. All participants in our study observed a rise in conflicts after the construction of the airport and the expressway. The following quotation by a man from Walsapugala captures the destitute situation of farmers in the district:

> Our hearts break when we see crop fields, where we put all our energy and money, destroyed by elephants. We don't eat or dress properly but sink into debt as our harvest has been destroyed. My father experienced this, now me, and if there are not solutions, our children will bear the same fate.[13]

The burden, however, is unevenly distributed. Interviewees revealed that poor farmers are the victims of elephants' attacks because they use solar-powered fences and firecrackers against the animals, provided to them by the DWC (DWC 2018; Prakash et al. 2020). In contrast, owners of agribusinesses were perceived to be unscathed, as they set up illegal electric fences to prevent elephants from crop raiding. These fences have further obstructed elephants' movements, thus increasing HEC and potentially causing the electrocution of elephants. Research has found that the lack of effectiveness and appropriateness of HEC mitigation methods and the development-induced land encroachment of wildlife habitats have escalated HEC in Hambantota (Prakash et al. 2020).

The safety and security of rural families has been challenged by elephant attacks. Many interview partners reported living with the fear of losing their lives. Some farmers and their families had spent nights in trees or in tree huts after elephants had damaged their houses. A female farmer from Ihala Andarawewa who engaged in the fast-unto-death campaign depicted the reality that they experienced as follows:

> Nowadays elephants are visible from 2 to 3 pm. If elephants remain on the road in the morning children can't go to school. And coming back home alone is unsafe. . . . [A]t night, we cannot seek help for health emergencies as elephants roam then. Husbands or male relatives have to guard crops all the time. We have endured the fear of losing life. . . . So, I sat for the fast-unto-death for 3 days to get [political] answers.[14]

Research (Fernando et al. 2011, 2020) shows that men are more likely to be killed by elephants than women and that most attacks occur at night. This might be because men are more likely to confront elephants at night on the fields, while women and girls are at home (Fernando et al. 2011, 2020; Prakash et al. 2020). Furthermore, the loss of a breadwinner is a tragic social and economic loss for households, which may push families into poverty (Prakash et al. 2020). Yet the compensation paid by the state for a human death caused by an elephant is as low as LKR 500,000 (ca. € 2,105) (DWC 2018: 16–24). What is clear from the analysis of these accounts is that the protests by the farmers were apolitical in the sense that they were neither linked to party programmes or political followership nor supported by external NGOs. Instead, community members who engaged in the protest were motivated by their prolonged hardships, including the fear for their lives. They requested activities and solutions from public authorities and politicians in power.

Restoring environmental justice through the elephant reserve

As the escalation of HEC has become a critical environmental concern in public debates, the local communities understand the need for an elephant reserve[15] in Hambantota to restore the ecological equilibrium. A female farmer from Walsapugala explained the following to us:

> Elephants need more food, water, and environment for mating and [raising] their calves. We did not ask for lands just for ourselves. Elephants are a valuable asset. . . . I feel terrible when I see a death of an elephant. . . . [T]hey should live their life freely in the woods, and it is their land too . . . but humans are also suffering.[16]

The sadness expressed by the woman over the death of an elephant resonates with Buddhist teachings on 'no harm for animals', which is embedded in these predominantly Sinhala Buddhist communities. Anger at human deaths caused by elephants is usually directed at government authorities, such as the Mahaweli Authority of Sri Lanka (MASL),[17] which oversees rural settlements, farms, and irrigation sites.

Interestingly, the second farmer leader from Walsapugala blamed himself for not foreseeing the human–wildlife implications prior to airport construction. Politicians romanticised how the international airport would improve rural people's lives by drafting an urban future for the district. Interview partners described the situation as them 'liv[ing] in a sweet dream painted by politicians'. In addition, interview partners perceived that there was no communication and coordination by environmental organisations with local communities on the predictable environmental damages. One interview partner commented that farmers' organisations could have resisted the decision of selecting Mattala as site for the new airport if they had been aware of and understood the situation correctly. In all, the communities' acknowledgement of wild elephants' right to exist in their neighbourhood reveals the building blocks needed to make interventions and policies towards the co-existence of elephants and humans.

Illegal land grabbing

For farmers in Hambantota, the elephant reserve not only is a way to mitigate HEC but also provides a way to prevent land grabbing. The illegal land encroachment in the proposed reserve has increased wildlife conflicts due to the reduction of forest areas for wild fodder, the fragmentation of elephant ranges, and the blocking of their pathways. For example, the elephant corridors from the MER to

Bundala, Gonaruwa, and Burutha Kanda have already been deforested, leaving less resources for elephants (Daily Mirror 2020, 2021).

Although land rights in the proposed elephant reserve belong to MASL and the Department of Forest Conservation (DFC), and are thus public land, interview partners blamed authorities for turning a blind eye to land grabbing for large-scale cultivation, illegal sand mining, and stone quarries. Interview partners thought that those who were performing land grabbing were the 'big and powerful', referring to alliances with government politicians and officials. One male farmer suggested to us that there was an unequal treatment to land users disadvantaging the poor farmers:

> Although the government say they care about agriculture, it is not about poor farmers like us. Those who have power and money have taken large plots of lands like 1 km^2 or 2 km^2 from the proposed elephant reserve's site. They now have deeds[18] as some politicians support these activities. It is illegal. But if we go to take even a piece of firewood, we have to face legal charges.[19]

There are allegations against MASL for releasing lands under its purview and conditioning the release of land for the elephant reserve (Daily Mirror 2021). Media (Daily Mirror 2021) have reported that the reserved forest lands have been cleared for various activities, such as for banana cultivation (8.1 km^2), private entities (24.3 km^2), and solar power projects (3.64 km^2).

Furthermore, interview partners claimed that large-scale cultivation in the elephant reserve exploits natural water resources. As the cultivators pumped water from lakes, for example, from Andarawewa, the water flow to their cultivations and for wildlife was drastically reduced. This large-scale resource grabbing is detrimental to both wildlife and poor farmers, who depend on the natural water supply (The Island 2021). Moreover, the increasing land value in the elephant reserve drives illegal encroachment there. A farmer commented 'the best land for future is in the elephant reserve', as the reserve is located in close proximity to the airport, the Hambantota Port, the Sooriyawewa cricket stadium, and the Southern Expressway. These infrastructures added value to the land. As a result, land grabbing, backed by former and current government politicians, increased in the reserve, leaving a small share of land and natural resources for elephants and poor farmers (Daily Mirror 2020).

Discussion and conclusion

The Walsapugala protest was a community call for more equitable sharing of resources and minimisation of the negative outcomes from airport construction. The study revealed that the implications from the intensification of HEC are unequally borne by rural poor farmers and their families. The local rural population was further excluded from the decision-making on development options for their area. In the inception, what the airport meant for these local communities were new employment opportunities for their younger generation, better access to markets when the district was to become an urban, commercial hub, and improved living standards for the villagers of Hambantota. In reality, Mattala Airport has failed to keep any of these promises. Furthermore, these communities have been prevented from implementing solutions, such as the elephant reserve, to mitigate long-term negative impacts from airport construction.

The protest focused on the gazetting of the elephant reserve and the sanctioning of land grabbing. It is important to note that farmers did not contest the airport or aviation development. Neither the airport as a place of employment nor new mobility options, which could eventually have been beneficial to them, became part of their livelihood; instead, Mattala Airport was used as a tool for the political glorification of the Mahinda Rajapaksa regime. Those who benefited from the airport were the Chinese bank that provided the loan and the Chinese government, which was able to increase their footprint in Sri Lanka. Moreover, this chapter shows how closely today's HEC is related to the airport construction ten years ago.

The farmers' attitudes of sharing land with and ensuring sufficient land for elephants suggest the development of community awareness and participation in elephant conservation to some extent. This indicates a shift from people expecting the government and relevant authorities to bear the sole responsibility for elephant conservation and mitigating HEC. Studies and the national policy on elephant conservation and management have also highlighted the need to create awareness among stakeholders, including village communities, to establish human–elephant coexistence and the management of elephants both inside and outside protected areas through alternative approaches, such as elephant reserves, land use planning, and other community participatory approaches (DWC 2006; Fernando et al. 2011).

The communities who engage in agriculture, like those in Walsapugala, may assume that agriculture is no longer viable due to the lack of economic and social protection and perceived social status and that only large-scale infrastructure projects can act as the catalyst for their socioeconomic development. The popular political slogan 'development for the village' (*gamata sanwardhanaya*) also denotes that development is synonymous with an urban setting that is indicative of moving away from agriculture. Following the 1977 introduction of the free market economy in Sri Lanka and the establishment of the free trade zones, political leaders of the time pledged to the rural youth that the creation of urban employment would result in improved living conditions, witnessing a surge in youth migration from rural areas to cities. More recently, the Port City in the commercial capital Colombo was depicted as a turning point in the country's economy by ruling party political figures, despite concerns over the growing Chinese presence and influence in Sri Lanka. Seemingly distracted by such pledges, which are mostly asserted on political platforms, the communities are usually complacent about taking the initiative in their own capacity to mitigate the environmental impact of such projects while shying away from urging the authorities to do so. In such a context, this chapter argues for making agriculture more attractive to the youth so that they might engage in sustainable livelihoods without leaving their communities.

Having experienced first-hand the severe ecological damage caused by Mattala Airport, the community-led non-partisan protest initiated by the community in Walsapugala sought the protection of elephants under the banner of 'people's voices from Walawe' (*walawen nagena jana ghosha*). While the community continued the protest for nearly three months, the state-owned media, unsurprisingly, refrained from covering the protest. In contrast, a handful of privately owned television and print media that are deemed anti-government continued to report about the protest. The leaders of the protest were later seen in the media unleashing scathing attacks on the government for its criticism of and inaction over the demands of the protest, albeit anti-government rhetoric was not dominant at the outset of the protest. Exploiting the Sinhalese community's prevalent negative attitudes towards the NGO community, the ruling party politicians responded by branding the people in Walsapugala as dependent on foreign-funded NGOs, possibly to shed a negative light on the protest and prevent the eruption of similar protests within the Sinhalese community.

On the other hand, the communities have little faith in officials and politicians. The ten-year delay in issuing the gazette of the MER, coupled with the issuing of the recent gazette containing an outdated map, has raised more questions than answers about the integrity of state mechanisms. While the people in Walsapugala slowly but steadily began to confront the realities of the environmental impact of Mattala Airport, successive governments neither listened to their grievances nor engaged with them. Predictably, the community in Walsapugala feels betrayed for having been abandoned by the current government, which they themselves supported during the 2019 presidential election. The government officials' supposed close connections with political cronies have plummeted public trust in state mechanisms, fostering the idea among the community that to be deceived both by politicians and officials is a driving force of the Sri Lankan political landscape. To compound matters, the Fauna and Flora Protection Ordinance Number 2 of 1937 is woefully inadequate to meet present-day needs and lacks protection for elephant corridors, according to most law experts. In such context, issuing the gazette of the MER alone is unlikely to be the panacea for all environmental ills faced by the community in Walsapugala.

In May 2013, an online petition urging the authorities to stop destroying the wildlife habitats in and around Mattala gained momentum. Predictably, this online petition never captured the attention of the community in Walsapugala, partly because the people had trust in the government at that time, given its development-centric approach being carried out close to their homes, especially within the electoral heartland of the Rajapaksa family. Nearly eight years later, in 2021, criticism of Mattala Airport came from the supposed beneficiaries themselves, who belatedly realised how the airport failed to fulfil their economic and other aspirations, having been built in the midst of a wildlife habitat where there is no air passenger traffic. Ironically, the airport was built while Mahinda Rajapaksa was the president, and the recent protest emerged while he is currently serving as the prime minister and one of his siblings, Gotabaya Rajapaksa, was the president, in addition to several other family members, including son of Mahinda Rajapaksa Namal Rajapaksa, holding ministerial positions in the administration. Without doubt, the protest was a bitter pill for the incumbent Rajapaksa government to swallow, as its image has been tainted in the face of the recent spate of protests. While Mattala Airport was regarded as a badge of success of the Mahinda Rajapaksa administration, its successor, Sirisena-Wickremesinghe unity government, decided to store paddy at its premises. This is indicative of how megaprojects are left in abeyance as pet projects of the political predecessor, rather than considered to be national assets that could contribute to economic development. Again, during the current presidency of Gotabaya Rajapaksa, plans were underway to expand Mattala Airport, showing how development projects are inextricably linked with enhancing the personal image of political leaders. For detractors, however, Mattala Airport is a reminder of increasing the Chinese presence and influence in Sri Lanka, as the airport has failed to generate revenue to this day.

Not consulting various stakeholders, as well as the absence of a mechanism of public hearings prior to the commencement of development projects, seems to have exacerbated the current tensions between state actors and the community in Walsapugala. The refrain '*idam mankollaya*' (land grabbing) used by the communities during the protest is poignant, although the agent of the activity (grabbing) is not explicitly indicated. If the communities claim that state actors are behind alleged land and water grabbing, it is unlikely that the solutions suggested to them by the political dispensation will bear fruit. The land–water nexus highlights how scare resources are being exploited to achieve commercial agendas by individuals and groups with political backing. Ironically, Mattala Airport has exacerbated land issues instead of leading to an increase in land value, as pledged via political platforms.

The farmers' understanding of natural resource grabbing, both land and water, in the elephant reserve reflects the unequal power relationships between the various stakeholders. The weakest are the voiceless elephants and wildlife in comparison to the poor farmers who have a voice and can resist; powerful actors are the illegal cultivators, politicians, and government authorities, such as MASL and DFC. Although infrastructure-driven projects such as Mattala Airport claimed to channel development for all, the ground reality reveals the unequal distribution of it and leaving the marginalised human populations in a further dire situation. In other words, the airport, as Bryant and Bailey (1997) argue, has exacerbated existing socioeconomic and political inequalities and altered power relationships in favour of the rich and powerful. Likewise, the silence from the state towards environmental protection suggests state-driven capitalism. In particular, the post-war infrastructure development, including the international airport by the Rajapaksa regime, provides evidence for the trend of prioritising industrialisation and corporate agribusiness over the agricultural livelihood of small-scale farmers, despite agriculture being the backbone of Hambantota.

A striking revelation from this case study was that the participation of minorities, Indigenous communities, university students, and the general public was either minimally present in or absent from this protest. Rather, this is an isolated, fragmented, localised struggle against environmental concerns by agrarian communities, with minimal support from non-farming groups. It is also interesting that international non-profit organisations had no involvement in assisting these farmers. Thus, future research may look into the causes behind the absence of island-wide activism against environmental depredation in Sri Lanka.

In conclusion, instead of viewing the constructed dichotomy of human and nature through an uncritical lens, the findings of this Mattala Airport case study identify the need to look at environmental issues through social, political, and economic paradigms. Through this approach, the case study unravelled complex dynamics, both visible and invisible on the airport land. While the protracted farmers' protest was not politically driven, the powerful actor, here the government, seeks exploitative control over natural resources. That said, further research is needed on these important issues as events unfold.

Acknowledgements

The authors would like to thank Irit Ittner and Sneha Sharma for their invaluable editorial input into this chapter. In addition, a special thanks goes to the interview partners in Walsapugala, as well as Dr Ewen Arnolds, Ken Kawasaki, and Visaka Kawasaki.

Notes

1 The national parks of Bundala, Udawalawa, Yala, and Lunugamwehera are located in the Hambantota district.
2 After decades of ethnic tensions between the Sinhalese and the ethnic Tamils, the infamous Black July anti-Tamil pogrom of 1983 led to war between the government armed forces and the LTTE that sought to establish an independent Tamil state (Eelam) in northern and eastern Sri Lanka, where the fighting often took place. In 2009, the Mahinda Rajapaksa administration declared victory over the LTTE, after most of their high-profile leaders were killed by the government forces.
3 In 2011, the Mahinda Rajapaksa administration made an unsuccessful bid to stage the 2018 Commonwealth Games in Hambantota, in an attempt to enhance the image of this rapidly transforming city.
4 The construction of Mattala Airport has cost US$247.7 million, while a loan agreement of US$209 million was entered into in 2009 between the Democratic Socialist Republic of Sri Lanka and the People's Republic of China via the Exim Bank of China (AGD 2015).
5 The Magampura Ruhuna Heritage Museum was established during the Mahinda Rajapaksa administration under the Department of National Museums, showcasing the cultural and natural heritage of the region of Ruhuna (Southern Sri Lanka).
6 In addition, the international cricket stadium in Hambantota was named after Mahinda Rajapaksa. The first two letters of the acronym of the international convention centre in Hambantota (MRICC) are also an implicit reference to the initials of Mahinda Rajapaksa (MR).
7 In 2018, President Sirisena appointed a five-member presidential commission to probe into allegations of fraud and corruption in SriLankan Airlines and Mihin Air, a low-cost carrier founded in 2006 during the Mahinda Rajapaksa regime that later ceased operations in 2016.
8 The EIA is enacted under the National Environmental Act No. 47 of 1980.
9 The main method to mitigate HEC and elephant conservation used in Sri Lanka is confining elephants to protected areas, such as national parks, by the Department of Wildlife Conservation (Fernando et al. 2011). However, it has been ineffective, as national

parks lose the carrying capacity to support elephants continually and impose detrimental effects on elephants (Fernando et al. 2011, 2021). Given the context of 70% of people residing in elephant ranges in Sri Lanka, in which the HEC occurs in overlapping areas, in situ management such as MER has been recommended (Fernando et al. 2021).
10 The 'fast unto death' means abstaining from food and drink till the demand is met politically or socially, failing which prepared for end.
11 Transcription of telephone interviews, 6 June 2021.
12 In February 2021, the Pottuvil-to-Poligandy protest marked a watershed event in the political landscape, during which the Tamils and Muslims rallied together to draw attention to issues that remain largely unaddressed in post-conflict afterlife. In addition, the Indian private funding for the Colombo Port's East Container Terminal (ECT) had to be withdrawn in response to trade union–led protests during the same period. The government's ad hoc ban on chemical fertiliser triggered farmer protests across the country, while the fishing community also staged demonstrations demanding compensation for the catastrophe caused by the sunken ship *Xpress Pearl*. In March, a protest was staged at Hingurakgoda in north-central Sri Lanka by the Collective of Women Victims of Micro Loans. This protest received minimal coverage by the mainstream media. In July 2021, many human rights advocates criticised the current administration for exploiting quarantine regulation to enforce quarantine on a group of teachers who protested over salary anomalies.
13 Transcription of telephone interview, 7 June 2021.
14 Transcription of telephone interview, 7 June 2022.
15 Interview partners were aware of the proposal for MER. Some of them joined the meeting of the district development committee in Hambantota, which approved it in 2012.
16 Transcription of telephone interview, 8 June 2021.
17 MASL was enacted under the Mahaweli Authority of Sri Lanka Act 1979 (No. 23 of 1979): 'It is responsible for the Mahaweli Ganga Development Scheme, provide for the establishment of Corporations to assist in such implementation and to provide for matters connected or incidental.' MASL has power to make decision on the optimisation of agricultural productivity and employment potential, as well as to improve the welfare of communities in special areas under Mahaweli Ganga development. The Hambantota district is among the six areas under the purview of MASL. It holds land rights to those six areas.
18 Legal document that shows land ownership.
19 Transcription of telephone interview, 8 June 2021.

References and media

Attanayake, C. (2018). 'Mattala: Attracting business into a lonely airport'. *Institute of South Asian Studies Working Paper No. 314*. Singapore: University of Singapore.
Auditor General's Department (AGD) (2015). *Selection of Mattala as the alternative international airport of Sri Lanka and its operations*. Performance and Environmental Audit Report PES/PE/MA/2015/02. Colombo: AGD.
Bryant, R. L. and S. Bailey (1997). *Third world political ecology: An introduction*. London: Routledge.
Bryman, A. (2012). *Social research methods*. Oxford: Oxford University Press.
Camisani, P. (2018). 'Sri Lanka: A political ecology of socio-environmental conflicts and development projects'. *Sustainability Science*. https://doi.org/10.1007/s11625-018-0544-7
Colombo Gazette (2016). Troops deployed to clear wild animals from Mattala. Retrieved 15.06.2021. https://colombogazette.com/2016/03/18/troops-deployed-to-clear-wild-animals-from-mattala/
Colombo Gazette (2021). Hambantota farmers launch death fast. Retrieved 30.06.2021. https://colombogazette.com/2021/01/20/hambantota-farmers-launch-death-fast/

Daily Mirror (2020). MER for jumbos obstructed by 'land work': Two-legged 'white elephants' with crooked mindset threaten Hambantota wild elephants. Retrieved 10.07.2021. www.dailymirror.lk/opinion/MER-for-jumbos-obstructed-by-land-work-Two-legged-white-elephants-with-crooked-mindset-threaten/172-202263

Daily Mirror (2021). Delays in gazetting proposed Managed Elephant Reserve in Hambantota. Retrieved 15.06.2021. www.dailymirror.lk/plus/Delays-in-gazetting-proposed-Managed-Elephant-Reserve-in-Hambantota/352-207787

Delang, C. O. (2005). 'The political ecology of deforestation in Thailand'. *Geography* 90(3), 225–237.

Department of Census and Statistics (2019). *Sri Lanka labour force survey. Annual report.* Colombo: DCS.

Department of Census and Statistics (2012). *Census of population and housing.* Colombo: DCS.

Department of Government Printing (DGP) (2021). *Gazette of extraordinary No. 2222/62.* Colombo: DGP.

Department of Wildlife Conservation (2006). National policy for the management of the wild elephant in Sri Lanka. Colombo: DWC.

Department of Wildlife Conservation (2018). Annual performance report 2018, Department of Wildlife Conservation. Colombo: DWC.

Economy Next (2021). Two fasting farmers in Hambantota hospitalised: Protest campaign continues. Retrieved 10.07.2021. https://economynext.com/two-fasting-farmers-in-hambantota-hospitalised-protest-campaign-continues-78109

Fernando, P. (2015). 'Managing elephants in Sri Lanka. Where we are and where we need to be'. *Ceylon Journal of Science (Biological Sciences)* 44(1). https://doi.org/10.4038/cjsbs.v44i1.7336

Fernando, D. S. T. P., M. Galappaththi, and U. K. G. K. Padmalal (2020). 'An overview of the human-elephant conflict in Tissamaharamaya, Hambantota District, Sri Lanka'. *Taprobanica, The Journal of Asian Biodiversity* 9(2), 210–216.

Fernando, P., J. Jayewardene, T. Prasad, W. Hendavitharana, and I. Pastorini (2011). 'Current status of Asian elephants in Sri Lanka'. *Gajah* 35, 93–103.

Fernando, P., M. K. C. R. D. Silva, L. K. A. Jayasinghe, H. K. Janaka, and J. Pastorini (2021). 'First country-wide survey of the Endangered Asian elephant: Towards better conservation and management in Sri Lanka'. *Oryx* 55(1), 46–55.

Fernando, P., E. Wikramanayake, D. Weerakoon, L. K. A. Jayasinghe, M. Gunawardene, and H. K. Janaka (2005). 'Perceptions and patterns of human – elephant conflict in old and new settlements in Sri Lanka. Insights for mitigation and management'. *Biodiversity & Conservation* 14(10), 2465–2481.

Forbes (2016). The story behind the world's emptiest international airport. Retrieved 30.06.2021. www.forbes.com/sites/wadeshepard/2016/05/28/the-story-behind-the-worlds-emptiest-international-airport-sri-lankas-mattala-

The Hindu (2021). A farmers' struggle in Sri Lanka, siding with elephants. Retrieved 21.06.2021. www.thehindu.com/news/international/a-farmers-struggle-in-sri-lanka-siding-with-elephants/article33776331.ece

The Island (2020). Whither the proposed elephant reserve? Retrieved 13.06.2021. http://island.lk/whither-the-proposed-elephant-reserve/ (Accessed 13 June 2021).

Lankanewsweb (2021). 'Farmers of Walsapugala continue passive resistance despite gazette'. *Lankanewsweb.* Retrieved 10.07.2021. https://www.lankanewsweb.net/67-general-news/82391-farmers-of-walsapugala-continue-passive-resistance-despite-gazette

LankaTruth (2021). 'People in Walsapugala pretend to fast but sleep soundly – Chamal Rajapaksa'. *LankaTruth,* 4 April. Retrieved 13.11.2021. https://lankatruth.com/en/?p=13548

Newsfirst (2021a). Hambantota farmers continue Satyagraha. Retrieved 10.07.2021. www.newsfirst.lk/2021/01/25/hambantota-farmers-continue-satyagraha/

Newsfirst (2021b). Protest on farmers, environment issues continue. www.newsfirst.lk/2021/03/28/protest-on-farmers-environment-issues-continue/

Newsfirst (2021c). 'Walsapugala farmers protest for 15th consecutive day'. *Sri Lanka News: Newsfirst*. Retrieved 30.06.2021. www.newsfirst.lk/2021/02/01/walsapugala-farmers-protest-for-15th-consecutive-day/

Newsfirst (2021d). 'Walsapugala farmers protest obstructing Kataragama: Colombo Rd'. *Sri Lanka News*. Retrieved 30.06.2021. www.newsfirst.lk/2021/03/10/walsapugala-farmers-protest-obstructing-kataragama-colombo-rd/

Peiris, V. (2019). 'Evaluation of post impacts of second international airport in Sri Lanka with reference to the environmental impact assessment process'. *International Journal of Multidisciplinary Research Review* 4, 76–83.

Prakash, S. L., A. W. Wijeratne, and P. Fernando (2020). 'Human-elephant conflict in Sri Lanka: Patterns and extent'. *Gajah* 51, 16–25.

Robbins, P. (2012). *Political ecology: A critical introduction*. 2nd Edition. Malden: Blackwell.

Sunday Times (2014). Religious beliefs save Mattala peacocks, but threat persists. Retrieved 15.06.2024. https://www.sundaytimes.lk/140119/news/religious-beliefs-save-mattala-peacocks-but-threat-persists-80589.html

Sunday Times Plus (2010). Wildlife in flight! Retrieved 14.06.2021. www.sundaytimes.lk/100822/Plus/plus_01.html

4 A critical review of airport land contestations in India

Sneha Sharma

Introduction: India's flying aspirations

> In the 75 years of civil aviation we have got to 75 airports. . . . [T]he good news is with UDAN [planning scheme, see below] we have added 31 new airports to the aviation network, we were able to design a market friendly and innovative scheme that made it attractive for airlines to go into the underserved market. . . . [W]e would like to add another 100 airports in the next 2–3 years.[1]

India's political independence in 1947 was followed by a sharp increase in airport construction, but the country's recent infrastructure-led economic development has led to an airport rush. As the preceding quote by the minister of aviation from 2017 illustrates, India is planning to construct 100 new airports by 2024. The International Air Transport Association predicts that India will become the world's third largest aviation market by 2030 (IATA 2018). A slew of airport projects have thus been initiated under a Regional Connectivity Scheme (RCS) that aims to connect larger cities and small towns within the air network across various Indian federal states. They include greenfield projects, rehabilitation initiatives, and modernisation efforts to expand and upgrade existing airports (see Section 2). However, these plans face multiple challenges as airport development primarily centres on land use change in the name of public interest. First, the Airport Authority of India (AAI) reports that there are around 18 under-served and 406 un-served airports lying idle and incurring costs.[2] The reasons for this vary from the airports being situated at remote locations to a lack of a national aviation policy. Secondly, airport infrastructures have overarching impacts across geographies as they are embedded in global and local networks. Airport development has faced global opposition and resistance from local communities, activists, and environmental regulators due to the social and ecological disruption they cause.[3] Prominent cases in India include the airports in Navi Mumbai, Pune, and Vizag, where residents protested against land use changes due to the greenfield development of airports. Similarly, a newly planned port township project threatened the autochthonous population of the Nicobar Islands in the Indian Ocean (30.01.2022 *The Print*). In 2017, plans for the Bhogapuram airport in Hyderabad faced protests when 376 families were

DOI: 10.4324/9781003494966-4

instructed to vacate their land, but the state controlled the agitation using police force and revised the compensation package they would receive (30.01.2022 *The Hindu*). Contentions have primarily precipitated around land values, set by the respective administrations, that determine the compensation value for project-affected citizens. Airport conflicts have intensified recently, affecting communities such as tea workers in Silchar (27.08.2023 Pratidin) Assam; Dalit and landless communities in Puri, Orissa; and small farmers in Surat, Gujarat (09.07.2021 *News Click*).

A quick review of the existing literature on aviation in India reveals a gap in studies that engage with the social aspects of nationwide conflicts and tensions due to airport expansion. A limited amount of research exists in the form of individual case studies and local newspaper sources (Jain & Watve 2020; Nair 2020; Nielsen 2015, 2018; Nielsen & Da Silva 2017). This chapter addresses this gap by analysing the nature of airport-related conflicts using existing literature sources. It aims to investigate ongoing tendencies in Indian land-related conflicts by zooming in on the territorial embeddedness of airports, that is, their local implications on communities and the environment. With this interest in mind, I present case studies drawn from reported conflicts across the country to answer the question: What are some of the emerging trends and patterns of land conflicts around airport projects in India? In addition, I analyse the social and environmental conflicts emerging around two study sites: Jewar in Uttar Pradesh and Dehradun in Uttarakhand. I undertake a detailed review of these cases because they remain insufficiently researched. These sites are critical for local communities and the environment, and they can inform planning strategies and conflict mitigation for future projects.

Political ecology has influenced and inspired this review by providing broad contours for case selection and analysis. This approach has enabled a layered, scalar, and contextual analysis of power relations around environmental conflicts (Blaikie & Brookfield 1987; Bryant & Bailey 1997; Ribot & Peluso 2003). Along these lines, airport land conflicts are understood as power struggles between actors over their access to material and social resources. Based on the findings, I argue that airports are a land question, that is, they are rooted in territorial and social entanglements that may enter into tensions with the larger development rationale at the national level.

The two cases bring together projects of different scales and politico-economic importance. Jewar is being built from scratch and is critical for northern India's economic growth, overtaking New Delhi airport, while Dehradun is a relatively small, tourism-focused expansion project striving to meet international standards. Both the airports are geopolitically significant. The land acquisition for Dehradun has been justified due to its proximity to the international border with China, while Jewar is planned to be connected to a Defence Industrial Corridor. Airport construction and rehabilitation faced immense opposition in these selected locations, and both projects are stuck in environmental battles that have led to judicial intervention and delays in implementation.

The cases studies were selected based on a preliminary literature review for the availability of recorded history and reliable data. The data sources for the analysis stem from secondary literature, which were available in either print or digital form. They include newspaper articles, grey literatures, academic literature in journals, books, and information made available by anti-aviation networks between 2000 and 2022. The data was then systematically analysed using two broad themes: (1) the role of development and (2) community protests. The cases are not compared with each other to avoid a generalising bias, but key insights from each are brought together to comment on the emerging trajectory of airport development in India.

The chapter begins by providing an overview of post-liberalisation policies and the privatisation of the Indian aviation industry from the 1990s. I then enumerate the debates around land acquisition, dispossession, and the role of the state, which provide a framework to analyse the cases. This is followed by detailed descriptions of the Jewar and Dehradun cases. I will demonstrate how utopian visions of development and conflict management are used to market the projects to the public and justify large-scale land acquisition for private profit and non-aeronautical uses.

Indian aviation after independence: a runway for investments

The aviation industry significantly contributes to India's global aspirations. It is seen as a jewel for regional growth. Unlike other transportation networks, such as railways and roads, airports provide dual functions that are essential for globalisation. On one hand, they are key pillars of the material infrastructure that caters to local economies. On the other, they enable freedom of mobility, trade, and resource flows across national boundaries. In the early 1980s, the Indian aviation industry worked towards making world-class international airports, but the current focus is on aggressively improving regional connectivity by reducing travel time. Airports are imagined to be prime infrastructure to attract other, non-aeronautical economic developments, such as urban services, tourism, jobs, and foreign direct investments. Furthermore, they influence social, cultural, and political structures. Cargo and freight flights are in high demand due to the explosion of online retail chains such as Amazon.

India's current air connectivity aspirations are set amidst the political environment of the 'Make in India' program announced by Prime Minister Narendra Modi. It aims to turn India into the world's new export hub. Dedicated investment hubs that will be connected through freight networks constituting rail, water, and air transport have been planned for this purpose. At the same time, the government has identified foreign investments as being extremely critical to develop local businesses. This dependence on infrastructural development is not new; India has been on a growth-led trajectory since the 1990s' adoption of neoliberal reforms (Kohli 2012). During this time period, the idea of 'development' was predominantly shaped by spectacular world-class infrastructure, the use of smart technology, and the growth of metropolitan cities. The aviation ministry saw immense potential in this economic environment to shape air transport into a competitive market with world-class infrastructure (Bansal et al. 2008). India adopted an open-sky policy in 1990 to encourage private airlines to operate anywhere across the country.[4] The policy was critical for the sector's development. It welcomed private players to provide air transport services by easing red tape, reducing taxation, and enacting reforms for airport construction. India passed the Air Corporations (Transfer of Undertaking and Repeals) Act 1994 to move away from the monopoly of a single national carrier and to allow economic institutions to be set up for a modern aviation industry operated by market mechanisms.

The National Civil Aviation Policy (NCAP) was formed in 2016 with the aim of enhancing regional connectivity by providing fiscal support and reviving airports' infrastructural improvements. It was believed that these two factors would draw in other benefits, such as employment, businesses, or real estate. Under the Regional Connectivity Scheme, another program called UDAN (Ude Desh ka Aam Nagrik; translated from Hindi as 'letting the common citizen of the country fly') was launched. Middle- and lower-income groups are the new target consumers for the aviation industry, which is now drawing on the country's demographics and the increasing disposable income of its service class.[5] Existing airports, particularly those situated in metropolitan cities, are under strain as domestic air passenger traffic more than doubled from 70 million in FY 2015 to 141 million in FY 2020.[6] India's dedicated ministry for civil aviation[7] believes that benefits from the sector can be derived by focusing on increasing the air traffic volume and erecting more airports.

Special provisions and exemptions have been made to allow for 100% foreign direct investment for airport construction (Rajan et al. 2009). Both state- and central-level policies provide huge concessions to private operators through instruments such as Viability Gap Funding, low service tax, and code-sharing schemes.[8] The construction of airports is largely supported by the state, which continues its entrepreneurial role of making land available at minimum costs to mostly foreign-funded investors. It is interesting to note that according to the NCAP policy, federal states are also liable to provide additional free land to develop hub-based transport networks (called multi-modal systems) that connect the airport with its hinterlands. Federal governments have to provide electricity, water, and other provisions to ensure the smooth functioning of this utility. The Indian government has undertaken a financing cycle of promoting greenfield megaprojects, financing new infrastructures, and improving existing airports and airlines by handing them over to private investors.[9] The NCAP allows private companies to apply for special tax rebates and concessions. This gives them the leeway to defy regulations, minimise environmental and social costs, and access cheap land in the name of development.

The Regional Connectivity Scheme is closely influenced by Western models of airport-oriented development, like the Aerotropolis (Kasarda & Lindsay 2011).[10] Another Western model that has impacted India's infrastructural development is the concept of 'greenfield' development, which encourages the development of identified agricultural or forest land parcels into new infrastructure. Kennedy and Sood posit that greenfield urbanisation is rooted in the utopian notion of 'tabula rasa', that is, 'to start from a clean slate' (2016: 43) to transform pristine, non-developed[11] land into urban frontiers. They argue that this form of development is not new for India and is what gave birth to new cities in the colonial era.

Theoretical inspirations

With traditions rooted in agrarian land and resource conflicts, political ecology centres on power and politics. It offers an appropriate lens to explain environmental conflicts (Escobar 2006; Martinez-Alier et al. 2016; Peet & Watts 1996; Perreault et al. 2015; Ribot & Peluso 2003; Peluso & Watts 2001; Zimmer 2010) and calls for an actor-oriented approach that focuses on the processes, that is, actors, institutions, and power relations, through which control over land (resources) is exercised. Turner (2004) argues that this complex interaction of contexts and actors may not be restricted to resource conflicts, but it may incorporate struggles over ethnicity, identity, gender, race, caste, etc. Contestations are primarily based around land transfer, the environment, and the displacement of native populations (Jewitt 2008). Airport development is embedded within a historical, colonial pattern of land acquisition as well as a contemporary, capital-driven logic of development. The state plays the role of facilitator and leverages its power of eminent domain to gain control of available land and transfer it primarily to private players (Shiva et al. 2011).

The 'eminent domain' is a doctrine from the colonial law. Through it, the state can 'overrule any dissent for land acquisition for public purpose' (Sampat 2013: 40). It is a legal power held by the state to acquire any private land for public use. Sampat argues that this doctrine has found an expression in current laws that regulate land acquisition. Another concept employed by the Indian developmental state, terra nullis, traces its origins to Roman law. It dictates that any empty land can be acquired by state force. Both these principles worked together in post-colonial times to displace landowners and sever their rights and resources. Defining public purpose is contentious. Following an amendment to expand the definition of the term 'public purposes', the Indian state's right to forceful control of land for 'public purposes' now includes projects by private parties (Sampat 2013).

Neoliberal projects tend to urbanise by erasing existing social histories. Land conflicts in India are often considered to be land grabs (Levien 2018) or land wars that politicise the acquisition of scarce land parcels. Transformation is brought about by breaking, demolishing, and displacing local people along with their practices, identities, memories, and sense of belonging. The lands acquired are not only empty spaces but also have social, cultural, and ecological value for both communities and wildlife. Jahnavi and Satpathy highlight that dispossession does not occur only from one's property, but also 'labour, social and cultural capital, and brutalities of identity hierarchies get perpetuated and extended' (2021: 5). Nielsen et al. rightly observe that 'land is back on the political agenda in India' (2020: 684). Reserved forest areas and commons such as grazing lands, fishing zones, and farmlands are being replaced by massive infrastructure projects, a phenomenon that Banerjee-Guha (2013) refers to as 'accumulation by dispossession'. With a pro-business government at the centre, Indian federal state governments are competing amongst themselves to provide land at cheap rates to such companies. Rana argues that modern land grabs have turned India into 'a "rentier state" [emphasis original] with an institutionalised nexus with the "power elite" comprising the politician, corporate/land

A critical review of airport land contestations in India 67

mafia and bureaucrat' (2020: 3). In the past few years, there have been frequent reports of locals protesting against land acquisition by the Indian state. Local communities have challenged the dominance of market mechanisms and the loss of livelihoods and ecological damage it creates (Sampat 2015; Sampat & Sunny 2016). Visible protests around proposed airport sites vary in scale and intensity; some were sporadic mobilisations that eventually dissolved, while others forced authorities to find alternative land (e.g. Dehradun) or subvert the project (e.g. Purandar, Pune) (07.08.2018 *The Times of India*; 06.01.2022c *The Times of India*). The next section details the conflicts in Jewar and Dehradun and describes local communities' struggles and the enduring tension between the centre and state relations.

Case studies

The greenfield airport in Jewar: farmers' protest for fair compensation

The largest planned greenfield airport in India is proposed to be Jewar in Gautam Budh Nagar District (Uttar Pradesh). It aims to ease the pressure on the existing New Delhi International Airport. Conceived in 2001, the Noida International Airport (NIA) will be built over 50 km² with six runways. It is designated to be India's most zero-carbon-efficient airport. The Zurich Airport International AG

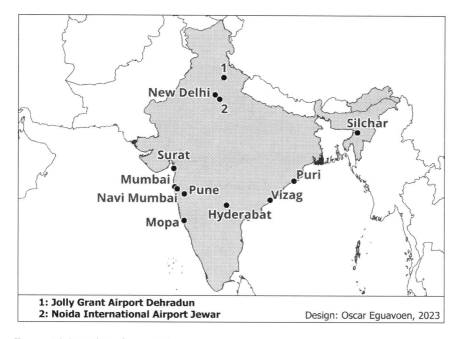

Figure 4.1 Location of case study sites.
Source: Design by Eguavoen.

has secured its operations for the next 40 years under a public–private partnership. Despite facing multiple political setbacks across government rulings due to its close location (25.11.2021 *Money Control*) to the existing New Delhi airport, its construction has been expedited under PM Modi's aggressive pro-business policy.

NIA is a critical component of infrastructure within the Yamuna Expressway Industrial Development Area (YEIDA), a newly planned city stretching over 165 km. YEIDA's dedicated Master Plan 2041 is based on the Aerotropolis model and includes special economic zones, industrial hubs, information technology parks, and residential spaces. The hinterland-based, multi-modal development project will connect another cross-country corridor, namely, the Delhi–Mumbai Corridor, with the Yamuna Expressway, a Rapid Rail Transport System, a metro corridor, and elevated road networks, such as the Western Peripheral Expressway. The Jewar airport is marketed as being the largest international logistics gateway (25.05.2021 *The Times of India*; 25.05.2022 *Business Today*).

Since the airport's foundation stone was laid in 2017, the media has repeatedly reproduced the image of a spectacular, sustainable, high-tech airport in the public domain. The preceding picture is one such vision of the development that appears in official state- and national-level narratives. Drawing from colonial land laws of the eminent domain, the state government will continue to acquire 13.65 km^2 in phase II, 13.18 km^2 in phase III, and 7.35 km^2 in phase IV of the airport construction (13.11.2020 *The New Indian Express*).

The ecologically rich airport land in Jewar consists of wetlands. Both social and ecological conflicts have emerged around the construction of the airport. Unlike other airport projects, the YEIDA has taken a proactive approach to preventing delays by pre-emptively seeking environmental clearances (ECs) that determine the approval of a project. The planning authorities seemed to have learned from the suspensions of other ECs, so they conducted an environmental impact assessment (EIA) in advance, and incorporated an afforestation plan to prevent potential stalling. During the EIA conducted by the Wildlife Institute of India, blackbuck and sarus crane habitats were found, which are categorised as vulnerable species in India (24.11.2021 *The Times of India*). The planning authorities drafted a conservation plan for the birds and fauna (14.08.2019 *Hindustan Times*). Estimated to cost €236 million, the conservation plan ultimately became a façade for greenwashing, as private companies were invited to contribute to it through their corporate social responsibility programmes.

Eight villages were directly affected by the land acquisition for phase I of the airport construction. According to the 2021 census, the *tehsil* (administrative unit) of Jewar consisted of both villages and urban settlements (Directorate of Census Operations Uttar Pradesh 2011). A social impact assessment (SIA)[12] was conducted by Gautam Buddha University, which estimated that 16,158 people would be displaced from these villages and urban settlements (2021: 7). While the project announcement and surveys initiated much speculation amongst the local community, the actual land acquisition began in June 2018, after the conclusion of a socioeconomic survey by Pricewaterhouse Coopers and the approval of the SIA. A rehabilitation and resettlement (R&R) program was prepared for compensation.[13]

The land conflict at the site primarily emerged around the compensation packages offered by YEIDA. The state government released €11.920 billion to compensate the farmers in the first phase of the project. The amount was fixed according to a rate determined by the authorities, which, as per the 'The Right to Fair Compensation and Transparency in Land Acquisition, Rehabilitation and Resettlement Act 2013' (RFCTLARR 2013) regulation, was two times the price of the circle rate applicable for urban areas (06.01.2022a Article 14).[14] Under RFCTLARR 2013, urban land can be acquired for only twice the circle rate, while rural lands require four times the circle rate as compensation. As the acquisition was underway, the farmers began claiming that they should receive four times the price of the circle rate, since their land came under the 'rural' category at the time of acquisition. The change of status of the villages was done in a piecemeal manner without informing the village councils (22.05.2022 *The Print*). This was partially done in 2015 by the Yamuna Authority, and the remaining villages were removed from the rural list in May 2018, just before the acquisition began. Following this, the panchayat, a crucial institution for rural governance in the acquired villages, was also dissolved. By May 2018, the farmers' dissent intensified when the locals were informed that the government had changed the status of the land parcels in Jewar from rural to urban. The farmers alleged that they were not made aware of the status change.

A farmer organisation, Jewar Kisan Sangharsh Samiti, translated from Hindi as 'Jewar Farmer's Struggle Union', led a series of peaceful demonstrations in February 2019 at the office of the district magistrate. Protesters claimed that the government had changed the official status of the land, yet it remained agricultural land in practice. They demanded a revision of the circle rate and held the government responsible for attempting to acquire land at cheap rates. They claimed that the YEIDA representatives had misinformed them and fraudulently obtained their consent and signatures. A writ petition was filed in the Allahabad High Court, demanding a four-time compensation increase to Rs 3,600/m2 (ca. €32/m2) and extra space to keep their cattle (08.04.2019 *The Times of India*). The High Court overruled their petitions for higher compensation. The agitation spread to the neighbouring villages for the remaining phases of the project.

The mobilisation of the farmers in Jewar was fragmented on many fronts, such as internal village and ethno-caste political disagreements. Many who owned large land parcels accepted the compensation and were ready to settle in other places. Land can only be acquired for public purposes under RFCTLARR 2013, but in the case of Jewar, the agricultural land was acquired by the state and leased out to private companies. Land acquisition for airports has considerably increased land prices and created a competitive real estate market even before the acquisition began. Due to long waiting periods for resettlement, many residents went back to cultivating their land. Farmers continue to protest across different pockets that await acquisition, while the state government has declared that the first phase of the acquisition is a conflict-free success.

Jewar's land is being acquired from farmers and agriculturalists and diverted towards large corporations who will gain control over its use. A significant observation is that the land is not the primary issue of contention in this scenario, but the rightful compensation package is the core concern for the farmers. Can this be framed as a resource conflict around competing land claims? I turn to political ecology to respond to this paradox. The farmers' approval to sell their land should not be seen in isolation from their struggle for access to compensatory resources. Their demands should be understood as material claims to access future resources. Turner (2004) argues that resource conflicts are not as straightforward as they might seem; they are superficial manifestations of embedded material and social relations.

Nair (2020) raises a similar puzzle in her study on the YEIDA acquisitions. She observes that farmers are ready to give up their land by claiming to be a part of market rationality and demanding a higher compensation. She contends that this can be explained through the concept of a moral economy, which allows the farmers to use land as a 'transactional asset' to safeguard their well-being by becoming part of market mechanisms. She warns that this is not a choice but a subsistence strategy that uses land to negotiate between urban imagination and the agrarian crisis.

The Jewar case points to a dissonance between the needs of social groups and the aspirations of the state, where power asymmetries in land acquisitions unequally fall on communities who might not even have the privilege to use air travel. The state uses various instruments and strategies to enable land transfer. First, a vision of development and unprecedented growth is discursively woven into the public narrative. This is backed by the current political regime's policies and geopolitical ambitions, along with a locally flourishing real estate market. Second, the affected communities are promised employment, training, and commercial land, which not only displaces their traditional occupations but also leaves them marginalised when these promises remain unfulfilled. Third, EIA and SIA are used as a politico-legal means to ensure the projects are approved, such as in the case of Mopa Airport (Aligbe & Prasad 2020). Often, the remedial measures, like nature restoration or employment generation, are not functionally implemented, which add to the symptomatic treatment of the ecological and social damage. Projects are broken down into phases, and land requirements are identified accordingly, but the impact assessment on local ecologies also becomes restricted to limited areas within each phase. In other words, studies are limited to isolated analysis. The fact that EIA and SIA are done separately is problematic. Instead, they should be brought together to better understand how co-existing environments and communities will be affected. This encourages determining a holistic picture of the projects' multi-layered, interrelated impacts.

Airport expansion in Dehradun: denotification of a wildlife reserve

Environmental protests have emerged against the planned expansion of Jolly Grant Airport in the Himalayan state of Uttarakhand. The airport was built in 1974 to improve domestic connectivity with Dehradun, a popular mountain destination visited by religious and leisure tourists.[15] The AAI announced the airport's world-class makeover in September 2020, with the aim to convert it into an international airport showcasing a high-tech design with an amalgamation with the local art and culture (Airports Authority of India 2021). It includes expanding the runway from 2,140 m to 2,765 m, constructing a new air traffic control unit, multi-level parking, and retail spaces within the airport premises. The expansion is planned across 0.8 km^2 of forest land and 15 km^2 of non-forest land. The airport is expected to provide easy access to the local tiger reserves to facilitate wildlife tourism. The state government has also allowed the construction of helipads at private resorts. Meanwhile, processes are already underway to identify a new greenfield airport in the state, which is planned to use over 4.45 km^2 of land.

Uttarakhand is covered by a recorded forest area of 71.05%, according to the Forest Survey of India.[16] The planned expansion impacts the only notified elephant reserve in the state, which has a population of over 2,000 elephants.[17] Elephant corridors are located within 1 km of the planned expansion area, and the Rajaji National Tiger Reserve is located within a 10 km radius. The local residents and farmers are directly dependent on these forests for produce, agriculture, and seasonal water sources.

In 2019, the state government of Uttarakhand approached the Ministry of Environment, Forest, and Climate Change (MOEF)[18] for approval to take over the forest land. It also ordered the National Wildlife Board to transfer a 'notified elephant reserve', the Shivalik Elephant Reserve (SER), to the AAI. In October 2020, the MOEF questioned the expansion plans and asked the state government to revisit its proposal while keeping in mind the impact on elephants. It further questioned the use of forest land for non-forest uses. Driven by inter-city competition and economic motives, the state government decided to change the official category of the elephant reserve to permanently bypass administrative hurdles in getting permission to use forest land. In November 2020, the Uttarakhand State Wildlife Advisory Board approved the request for de-notification of SER, which came as a shock to environmentalists. The state claimed that the SER was a notified zone, but it was not legally protected. The SER had become a notified elephant reserve in 2002 under the national program 'Project Elephants' that aimed to protect the endangered species. Environmentalists say that the surrounding area falls under the 'high conservation value' category, and construction will destabilise mountain slopes and increase wildlife–human conflicts (27.12.2020 *Bloomberg Quint*). This will directly cause elephants to enter human habitation when their habitats are destroyed. On 16 September 2021, an elephant entered the airport premises and walked on the runway by breaking a portion of the surrounding wall (16.09.2021 *The Times of India*).

This conflict exposes a tension in governance between the central and federal authorities. Land has to be made available by the state for any developmental activity, but in this case, the area is part of SER, that is, a central government notification that the state wishes to overrule. The MOEF once again sent a notice rejecting the justification offered by the state and insisted that they look for an alternate site. Local residents and activists protested and filed a public interest litigation in November 2020 (06.01.2022b *The Times of India*). The protests against the Dehradun airport are primarily led by young college students and local residents. The project has been politicised by the ruling party, who cite the strategic importance of the airport for defence, as the federal state shares borders with both Nepal and China. However, mobilisation has been fragmented when compared to the state's past ecological movements. The local farmers and inhabitants have been promised government jobs, and this has prevented a stronger mobilisation like the Chipko movement.[19] Several farmers hesitated to speak against the project, fearing loss of employment opportunity. The current upgrade[20] is expected to fell 9,745 trees like timber trees, sagoon, and khair, all of which are essential for carbon sequestration and ecological habitats.

The state has proposed mitigation plans, such as adequate fencing, elephant-resistant walls, water holes, and rescue gear, and it plans to carry out afforestation for the felled trees. The state's forest department has proposed a mitigation plan by claiming that the airport will be eco-friendly and utilise rainwater harvesting, solar power, sewage treatment, etc. Planting saplings takes time to be able to support biodiversity and wildlife; hence, it is not a sustainable alternative to destroying existing forests. Environmentalists have brought to light the state's incapability to finish afforestation measures for other projects, a case similar to the Mopa airport project, where land for afforestation was not available within the Goan federal state (04.06.2013 *The Economic Times*; 28.01.2021 *The Times of India*).

Several issues come to light in relation to the operation of compensatory afforestation, a tactic that is so extensively used by planning bodies to get projects approved.[21] Studies have reported a critical shortage of land banks available for compensatory afforestation to Indian state authorities. There are structural and institutional shortcomings that have backfired on the mandate for compulsory afforestation in projects. For instance, Saxena (2019) identifies that the nature of land that can be used for afforestation is vaguely defined as 'non-forest' and fails to clarify or address diverse issues. He demonstrates that the pressure to adhere to these have led to further land grabs from autochthonous and tribal communities who have varied communal arrangements and customary practices. It exaggerates the existing conflicts over legal land titles and common property.

The case foregrounds the idea that the impact of biodiversity loss can be mitigated through financial means by paying for afforestation. Through the impact assessments required for ECs, their post-project impacts can be closely estimated. Projects are given clearances in any case and on the assumption that the promised actions will be executed. Menon, Kohli, and Gupta (28.09.2017 *The Wire*) draw from a range of infrastructural projects to prove that EC monitoring is weak, and there are dire inconsistencies between the expected compliance and on-ground practices. It is not just airports but a string of other infrastructural activities, such as road construction for development of Char Dham Yatra (religious tourism), the construction of resorts, mining, etc., that have led to a loss of forest cover in the state. In 2013, Uttarakhand witnessed the country's most disastrous floods, which killed more than 5,700 people. A significant reason for this disaster was the massive scale of unplanned construction activity and flouting of environmental regulations by the tourism industry.

The case of Dehradun highlights an issue of land use change at the cost of rich ecological value, which is bound to disrupt human–wildlife relations. It illustrates the failure of state instruments to mitigate conflicts and reveals how compensatory measures to obtain ECs perpetuate land-related conflicts, as afforestation cannot replace an elephant corridor. The state government's approval for the denotification of SER is a means to accelerate other infrastructural projects in eco-sensitive areas, thus continuing ecological destruction and prioritising infrastructure-led expansion. This model of airport expansion directly contradicts the state government's claims of sustainability in aviation. Furthermore, it creates a tendency to supersede centrally implemented rules for nature protection. Menon, Kohli, and Gupta (28.09.2017 *The Wire*) have argued that state governments often tend to dilute central regulations, bypass ordinances, and refuse to return unused, acquired land, all of which highlight the tensions between federal states and the centre. The impacts of SER's de-notification will not be limited to the airport expansion, but it is bound to act as an official precedent for the future manoeuvring of regulations and institutions dealing with the loss of forest land.

Discussion and conclusion

Airport conflicts are complex issues that incorporate a neglect of local needs, politics, the state's structural deficiencies, hurdles in state–market coordination, poor implementation of rehabilitation policies, flouting of regulations, litigations, and hasty impact assessments. The contrast between the vision, functions, and planning of the two projects allows me to weave together insights on airport planning patterns and the governance of conflict management as undertaken by different stakeholders.

Each conflict follows its own trajectory, but a review of airport conflicts in India reveals several prominent attributes at play. I have divided this chapter into two sections: development aspirations and tools of displacement. The first addresses the visions of development and use of public interest to justify displacement and ecological destruction. The second discusses the structural issues around environmental conflicts and the use of compensatory and restorative measures to greenwash harmful consequences.

Utopian visions of development

Infrastructural projects are driven by a vision that justifies their irrefutable need. The diversion of land for megaprojects is often justified on the basis of population pressure, international competition, or technological advancement. There are many factors that drive state interest in airport development, such as connectivity, local trade networks, regional geopolitics, and inter-federal competition. The state releases land and other resources for private players in anticipation of the economic benefits that airports provide. Kennedy and Sood (2016) argue that state actors, in their bid to accelerate capital accumulation, depend on expropriate control over land and resources by a rule of exception. These exceptions can range from tweaking land-zoning rules and coastal protection regulations to amendments in industrial policies. All are done solely in favour of global capital. Anticipation of growth and risk-taking by the state is critical for such projects, since benefits are not always guaranteed. This was evident in Sri Lanka's case study of this book edition (Gunasekara and Senaratne 2024 – this book).

As evident in both case studies, visibility is generated through public discourse, and the detrimental impacts of these projects, such as ecological degradation, are whitewashed by grandiose visions of quantifiable economic growth. The stakeholders who drive and benefit from these megaprojects primarily consist of the state, private investors, middle and affluent classes, and politicians. With UDAN's objective to make the masses fly, it is clear that the lower-middle classes have been identified as air transport's new consumers. But in reality, most airport infrastructure is built for those with resources who cannot only purchase air tickets but also consume the luxury services and experiences they provide.

Apart from the attempts to package these airport aspirations in specific ways to the masses, a constant feature in both case studies is economic development. Both airports maintain that they will attract non-aeronautical utilities and industries through real estate and capital investment. Not just the airport area but the planned multi-modal transport infrastructures and connectivity with other national highways and corridors cater to a speculation in rising land prices due to investment by private capital, who buy or lease parcels of land for hotels and townships even before the airport construction has begun. Similar observations can be made in the cases of Pune (07.08.2018 *The Times of India*) and Vizag (18.03.2021 *News Click*), where the planning authorities and state governments witnessed a boom in the rentier land market due to the announcements of these projects. These visions, therefore, create a multiplier effect in the local land prices that can set in motion illicit and volatile markets (Sharma 2023). However, the preceding cases also demonstrate that visions and governance of the centre and federal states might stand in contrast with each other. Being at the forefront of acquiring land directly from owners, the state strives to maintain authority in land transactions of national importance while negotiating its position to compete with and secure control over incoming investments.

Public interest is another key aspect in launching these projects. They are portrayed as a necessity for the 'welfare' of larger sections of society. But on the contrary, the rhetoric is extended to non-aeronautical purposes and ancillary developments, such as adventure parks, shopping malls, parking spaces for cars, etc. Homogenous depictions of populations fail to point out who would be the actual beneficiaries of these projects. Scholars like Dhru (2010) have pointed out past efforts by policymakers and the Supreme Court to restrict the meaning of 'public purpose', but its interpretation was left up to the government. Moreover, the concept of public interest is reductive when it comes to ecology, because it excludes incorporating wildlife ecosystems in the planning of infrastructures, as in the case of Dehradun. The next section addresses the use of restorative and compensatory tools to mitigate environmental conflicts.

Strategies and instruments for land acquisition: environmental clearances, centre stage governance, and a piecemeal approach

The imperative for connectivity to boost economic growth does not only convert important agricultural land and dispossess farmers but is also a threat to the delicate ecological balance. Escobar contends that capital has entered a phase where nature is made part of its development agenda, that is, 'two old enemies, growth and the environment, are reconciled' (1996: 52). Economic growth is now packaged with promises of green futures. Concepts such as sustainable development and corporate social responsibility have become the rhetoric to continue the destruction of nature in return for tokenistic conversation efforts. Infrastructural projects, whether greenfield or not, treat land as a blank canvas that is bereft of ecological, social, and cultural values. Valuable agrarian and forest land is taken for non-aeronautical uses that require the destruction of forest cover, the diversion of rivers, and the blasting of hills.

In the past, when available land parcels were used for development projects, only a small section of the population actually benefited from its economic rewards. Funds invested in biodiversity conservation become the selling point, and these actions are lauded as capable of offsetting the significant ecological harm and social fragmentation caused by mega projects.

Another important insight is that the EIAs, SIAs, and ECs that are approved with some addendums for mitigation and conservation are a mere tokenism. Land acquisition is approved based on a central ministry-approved EC, which becomes a means to justify execution of projects. ECs are sought and granted based on conservation plans and afforestation (Menon & Kohli 2015), but environmental scientists have pointed out that the ecosystems that get destroyed during such projects cannot be replaced by afforestation efforts. In cases where the EC is challenged, the concerned planning agencies undertake revisions based on additional impact assessments. The established way to get through an EC is revisiting the problematic areas of the conservation plan. In any case, the impacts of faulty or hasty clearances are felt by local communities. Authorities seem to have learned well from a history of EC rejections. Therefore, like in the case of Jewar, EIAs and SIAs are undertaken in advance to ensure project approval. Underlying the restorative and conservationist measures of development projects is the idea that ecological damage can be compensated for and the need for development justifies a certain amount of ecological damage.

Secondly, a consistent strategy for airport development has been a piecemeal approach to land acquisition. When a project is broken down into phases, its impacts are also estimated in a fragmented manner. A segmented assessment, along with the complex involvement of multiple stakeholders, prevents a scalar, long-term investigation of the actual impact of land-related conflicts that might arise. Airport development across the country is not an isolated undertaking, but it forms part of a larger regional infrastructural rush in the country.

Lastly, the review of cases in India reiterates the primary contention in regulatory frameworks and decision-making, that is, the binary thinking between nature and culture. Conservation efforts and compensation packages reflect a narrow understanding of the interrelationships between nature and humans (in other words, development). This defines how land conflicts are shaped through capital-intensive projects and further deepens the boundaries between the social and material worlds by pitting one against the other in the name of development. These are only a few insights on contestations around airport land in India, but the chapter demonstrate the requirement for more in-depth research into policy gaps and the power relations that shape the politics of airport land struggles.

Acknowledgements

The research was conducted in the project titled 'Urban Villages by the Airport: Everyday Entanglements of Social-economic Extremes and Negotiations in Anticipation of Development-Induced Displacement in Mumbai and Abidjan', which was funded by the Fritz Thyssen Foundation (2020–2023, 10.20.2.003EL). I would like to thank Irit Ittner for the review of earlier versions, and Oscar Eguavoen for designing the map.

Notes

1. Quote by Mr Shri Jayant Sinha, Minister of State for Civil Aviation, at a panel discussion at the Confederation of Indian Industry Annual Session, 28–29 April 2017, at New Delhi. Available on YouTube: www.youtube.com/watch?v=zQKjrdG-mVI.
2. Data taken from Ministry of Civil Aviation website at www.aai.aero/en/rcsudan.
3. See *Environmental Justice Atlas* website at https://ejatlas.org/.
4. The policy was implemented initially for three years and then made permanent through the Aeronautical Information Circular AIQ No. 18/1992. For details, see report by International Civil Aviation Organization (2003).
5. See the official AAI website at www.aai.aero/en/rcsudan/rcs-udan-1/2268.
6. See website at www.aai.aero/en/content/airport-policy.
7. Infrastructural development needs institutions, strategies, and policies that align with the market, and the civil aviation policy caters to these needs.
8. Viability gap funding is a scheme where the central government provides 20% of the cost of total project budget to support a public–private partnership in infrastructural development. Code-sharing is a market agreement between airline companies through which they can sell partner airlines' tickets for sectors they do not cover. The government regulates code-share agreements.
9. In October 2021, the country's national carrier, Air India, was sold to the TATA group, a private company, as a disinvestment measure.
10. This form of spatial growth incorporates aviation and non-aviation-related activities connected via a road and/or rail network.
11. Here, the word *developed* indicates a land parcel on which some kind of construction has been carried out.
12. The RFCTLARR 2013 (Right to Fair Compensation and Transparency in Land Acquisition, Rehabilitation and Resettlement Act) was implemented in 2013, replacing the colonial Land Acquisition Act from 1894. The new Act aimed to ensure regulated, transparent, and fair land acquisition in the country. The law mandates that a social impact assessment be conducted. It ensures provisions for consent and public consultations in cases where land is compulsorily acquired from private holders for public purposes.

13 Along with this, the R&R package offered the farmers a possibility of one job per family and transport cost. A 50% developed plot would be provided proportionate to their acquired property.
14 Circle rate is the minimum price fixed by a government authority. It was fixed at Rs 2,300–2,500/m² (ca. $31–34/m2).
15 The state has three other airports, Bharkot, Pantnagar, and Naini Saini, which are either not for commercial use or have low capacities.
16 See India Forest Report 2027, Chapter on Forest and Tree Resources in States and Union Territories; Online Source https://fsi.nic.in/isfr2017/uttarakhand-isfr-2017.pdf.
17 Elephant census conducted by forest department. It stated that Uttarakhand has over 2,000 elephants, a 10% increase from 2017, from 1,839 to 2,026, as of June 2020.
18 Clearance from centre is necessary under the Forest Conservation Act 1980.
19 A collective mobilisation where locals opposed various instances of privatisation of forest land. It became a popular movement in the same state in the1970s.
20 An RTI revealed that during a previous expansion work of the airport in 2002, around 9,358 trees were felled.
21 In 2016, the GOI passed the Compensatory Afforestation Fund (CAF), through which it set up institutional measures to manage afforestation. The penalty paid by the project agency is collected by the state and deposited to a central authority, which then disburses them back to the state to carry out afforestation. Compensatory afforestation is mandatory for getting EC. Kohli and Menon (2016) note that monetisation of compensation has led to the collection of large funds, but these funds still lie unused due to lack of land available. This interrelated structural deficiency remains hidden, while afforestation is promised to seek land diversion. This defies the intentions of protecting the forests.

References

Aligbe, M. and R. Prasad (2020). '*The environmental assessment basis of the decision to build a new airport at Mopa in Goa, India*'. Project Report published on Researchgate. https://doi.org/10.13140/RG.2.2.31591.06560

Bailey, S. and R. Bryant (1997). *Third world political ecology*. London and New York: Routledge.

Banerjee-Guha, S. (2013). 'Accumulation and dispossession: Contradictions of growth and development in contemporary India'. *South Asia: Journal of South Asian Studies* 36(2), 165–179. https://doi.org/10.1080/00856401.2013.804026

Bansal, S. C., M. N. Khan, and V. R. Dutt (2008). 'Economic liberalisation and civil aviation industry'. *Economic and Political Weekly* 43(34), 71–76.

Blaikie, P. and H. Brookfield (1987). *Land degradation and society*. London: Routledge.

Dhru, K. A. (2010). *Acquisition of land for development projects in India: The road ahead*. Gujarat: Research Foundation for Governance in India.

Directorate of Census Operations Uttar Pradesh (2011). *District Census Handbook: Gautam Buddha Nagar. 10. Part XII-B*. Lucknow: Directorate of Census Operations, Government of India. http://censusindia.gov.in

Escobar, A. (1996). 'Constructing nature: Elements for a poststructural political eecology'. In *Liberation ecologies: Environment, development, social movements*. R. Peet and M. Watts (eds.). London and New York: Routledge, 46–68.

Escobar, A. (2006). 'Difference and conflict in the struggle over natural resources: A political ecology framework'. *Development* 49, 6–13. https://doi.org/10.1057/palgrave.development.1100267

Gautam Buddha University (2021). *Social impact assessment study: Noida international airport, report submitted for expert committee review*. Greater Noida: Gautam Buddha University.

Gunasekara, M. and D. Senaratne (2024). 'The rise of infrastructure-induced human – elephant conflict in Sri Lanka. A case study of Mattala Rajapaksa International Airport'. In *Contested airport land. Social-spatial transformation and environmental injustice in Asia and Africa*. I. Ittner, S. Sharma, I. Khambule, and H. Geschewski (eds.). Abingdon: Routledge.

International Air Transport Association (2018). 'Potential and challenges of Indian aviation'. *Press Release 50*. www.iata.org/en/pressroom/pr/2018-09-04-02/

International Civil Aviation Organization (2003). India's open skies policy on Air Cargo. www.icao.int/sustainability/CaseStudies/StatesReplies/India_En.pdf

Jahnavi, K. L. and S. Satpathy (2021). 'Unfolding the enigma of dispossession in India: An analysis of the discourse on land grabbing'. *Sociological Bulletin* 70(3), 331–348. https://doi.org/10.1177/00380229211018986

Jain, N. and A. Watve (2020). 'Contestation of environmental impact assessment for Greenfield Airport, Mopa, India through the lens of vulnerability'. *Indian Journal of Environmental Sciences* 24(1), 1–15.

Jewitt, S. (2008). 'Political ecology of Jharkhand conflicts'. *Asia Pacific Viewpoint* 49(1), 68–82. https://doi.org/10.1111/j.1467-8373.2008.00361.x

Kasarda, J. D. and G. Lindsay (2011). *Aerotropolis: The way we'll live next*. London: Allan Lane/Penguin Books.

Kennedy, L. and A. Sood (2016). 'Greenfield development as Tabula Rasa'. *Economic and Political Weekly* 17, 41–48.

Kohli, A. (2012). *Poverty amid plenty in the New India*. Cambridge: Cambridge University Press. https://doi.org/10.1017/CBO9781139015080

Kohli, K. and M. Menon (2016). 'Monetising the destruction of ecological landscapes'. *Current Conservation* 10(4), 14–17. www.currentconservation.org/monetising-the-destruction-of-ecological-landscapes/

Levien, M. (2018). *Dispossession without development: Land grabs in neoliberal India. Modern South Asia*. Oxford: Oxford University Press. https://doi.org/10.1093/oso/9780190859152.001.0001

Martinez-Alier, J., F. Demaria, L. Temper, and M. Walter (2016). 'Changing social metabolism and environmental conflicts in India and South America'. *Journal of Political Ecology* 23(1). https://doi.org/10.2458/v23i1.20252

Menon, M. and K. Kohli (2015). 'Environmental regulation in India: Moving 'forward' in the old direction'. *Economic and Political Weekly* 50, 20–23.

Nair, M. (2020). 'Land as a transactional asset: Moral economy and market logic in contested land acquisition in India'. *Development and Change* 51(6), 1511–1532.

Nielsen, K. B. (2015). 'Mopa Airport woes'. *Economic and Political Weekly* 50(25), 7–8.

Nielsen, K. B. (2018). *Land dispossession and everyday politics in rural Eastern India*. Anthem Press. https://doi.org/10.2307/j.ctt1zxsknd

Nielsen, K. B. and S. Da Silva (2017). 'Golden or green? Growth infrastructures and resistance in Goa'. In *Urban utopias*. T. Kuldova and M. A. Varghese (eds.). Cham: Springer, 53–73. https://doi.org/10.1007/978-3-319-47623-0_4

Nielsen, K. B., S. Sareen, and P. Oskarsson (2020). 'The politics of caste in India's new land wars'. *Journal of Contemporary Asia* 50(5), 684–695. https://doi.org/10.1080/00472336.2020.1728780

Peet, R. and M. Watts (1996). *Liberation ecologies: Environment, development, social movements*. London: Routledge.

Peluso, N. L. and M. Watts (2001). *Violent environments*. Ithaca, NY: Cornell University Press.

Perreault, T., G. Bridge, and J. McCarthy (2015). *The Routledge handbook of political ecology*. Abingdon: Routledge.

Rajan, A., S. S. Thillai, and S. Sidharth (2009). 'Public private partnership in Greenfield Airport development: A case of Cochin International Airport Limited'. In *Policy, management and finance of public-private partnerships*. A. Akintoye and M. Beck (eds.). Chichester: John Wiley & Sons.

Rana, S. (2020). 'Cartography of land acquisitions, "gated" development and the Bhu-Andolan that changed the discourse on Bhu-Swaraj'. In *ISA eSymposium for Sociology* 29, 1–18. www.isaportal.org/resources/resource/cartography-of-land-acquisitions-gated-development-and-the-bhu-andolan-that-changed-the-discourse-on-bhu-swaraj/

Ribot, J. C. and N. L. Peluso (2003). 'A theory of access'. *Rural Sociology* 68(2), 153–181. https://doi.org/10.1111/j.1549-0831.2003.tb00133.x

Sampat, P. (2013). 'Limits to absolute power: Eminent domain and the right to land in India'. *Economic and Political Weekly* 48(May), 40–52.

Sampat, P. (2015). 'The "goan impasse": Land rights and resistance to SEZs in Goa, India'. *The Journal of Peasant Studies* 42(April), 1–26. https://doi.org/10.1080/03066150.2015.1013098

Sampat, P. and S. Sunny (2016). 'Dholera and the myth of voluntary land pooling'. *Socio-Legal Review* 12(2), 1–17.

Saxena, K. B. (2019). 'Compensatory afforestation fund act and rules: Deforestation, tribal displacement and an alibi for legalised land grabbing'. *Social Change* 49(1), 23–40. https://doi.org/10.1177/0049085718821766

Sharma, S. (2023). 'Geographies of exclusion. Reproducing dispossession and erasure within a waste picker organization in Mumbai'. *International Journal of Urban and Regional Research* 47(5), 861–875.

Shiva, V., S. Jani, and S. M. Fontana (2011). *The Great Indian land grab*. New Delhi: Navdanya.

Turner, M. D. (2004). 'Political ecology and the moral dimensions of "resource conflicts": The case of farmer – herder conflicts in the Sahel'. *Political Geography* 23(7), 863–889. www.sciencedirect.com/science/article/pii/S0962629804000654

Zimmer, A. (2010). 'Urban political ecology: Theoretical concepts, challenges, and suggested future directions'. *Erdkunde* 64(4), 343–354. www.jstor.org/stable/25822107

Media sources

04.06.2013. 'Goa Govt oppressing farmers over land for Mopa Airport: Samiti'. *The Economic Times*, Press Trust of India. Retrieved 24.06.2022. https://economictimes.indiatimes.com/news/politics-and-nation/goa-govt-oppressing-farmers-over-land-for-mopa-airport-samiti/articleshow/20431202.cms?from=mdr

28.09.2017. 'In state-level changes to land laws, a return to land grabbing in development's name'. *The Wire*. Retrieved 15.04.2022. https://thewire.in/law/state-level-changes-land-laws-return-land-grabbing-developments-name

07.08.2018. 'Villagers block roads for four hours: Villagers block roads for four hours to protest against Purandar Airport'. *The Times of India*. Retrieved 12.11.2022. https://timesofindia.indiatimes.com/city/pune/villagers-block-roads-for-4-hours-to-protest-against-purandar-airport/articleshow/65298641.cms

08.04.2019. 'Allahabad HC rejects Jewar farm owners' writ petitions over airport project'. *The Times of India*. Retrieved 15.10.2020. https://timesofindia.indiatimes.com/city/noida/allahabad-hc-rejects-jewar-farm-owners-writ-petitions-over-airport-project/articleshow/68781051.cms

14.08.2019. 'Jewar Airport: Wildlife institute to prepare conservation plan for seeking environmental clearances'. *Hindustan Times*. Retrieved 03.12.2022. www.hindustantimes.com/cities/jewar-airport-wii-to-prepare-conservation-plan-for-seeking-environmental-clearances/story-ANiRzqTAdDtUBay5T4ieRO.html

13.11.2020. 'Jewar Airport project: Yeida seeks Yogi government go-ahead to begin land acquisition'. *The New Indian Express*. Retrieved 22.09.2020. www.newindianexpress.com/nation/2020/nov/13/jewar-airport-project-yeidaseeks-yogi-governmentgo-ahead-to-begin-land-acquisition-2223096.html

27.12.2020. 'When development marches through Himalayan forests'. *Bloomberg Quint*. Retrieved 16.04.2022. www.bloombergquint.com/shell.html

28.01.2021. '54,176 Trees cut at Mopa site, 500 replanted: Goa CM'. *The Times of India*. Retrieved 15.06.2022. https://timesofindia.indiatimes.com/city/goa/54176-trees-cut-at-mopa-site-500-replanted-cm/articleshow/80488795.cms

18.03.2021. 'Locals' issues ignored in clearance to new international airport in Andhra Pradesh?'. *News Click*. Retrieved 16.04.2021. www.newsclick.in/Locals-Issues-Ignored-Clearance-New-International-Airport-Andhra-Pradesh

25.05.2021. 'Warehouse, logistics hub may come up close to Noida Airport'. *The Times of India*. Retrieved 05.02.2022. https://timesofindia.indiatimes.com/city/noida/warehouse-logistics-hub-may-come-up-close-to-noida-airport/articleshow/82923123.cms

09.07.2021. 'Puri Airport: Dalit and landless communities continue struggle for land rights'. *News Click*. Retrieved 15.06.2022. www.newsclick.in/puri-airport-dalit-landless-communities-continue-struggle-land-rights

16.09.2021. 'Dehradun: Jumbo jets into runway, breaks down airport wall'. *The Times of India*. Retrieved 20.09.2021. https://timesofindia.indiatimes.com/city/dehradun/jumbo-jets-into-runway-breaks-down-airport-wall-at-exit/articleshow/86237982.cms

24.11.2021. 'The green challenge of building India's newest airport'. *The Times of India*. Retrieved 12.11.2022. https://timesofindia.indiatimes.com/india/the-green-challenge-of-building-indias-newest-airport/articleshow/87752682.cms

25.11.2021. 'Jewar Airport to be North India's logistics gateway; Noida, Ghaziabad Manufacturing hubs to get boost'. *Money Control*. Retrieved 22.11.2022. www.moneycontrol.com/news/business/real-estate/jewar-airport-to-become-logistics-gateway-of-north-india-manufacturing-hub-of-noida-greater-noida-ghaziabad-to-be-benefitted-7762371.html

06.01.2022a. 'Denied lawful compensation for land acquired for Noida international airport, farmers contest Modi's claims'. *Article 14*. Retrieved 22.02.2022. https://article-14.com/post/denied-lawful-compensation-for-land-acquired-for-noida-international-airport-farmers-contest-modi-s-claims--61d6535e80940

06.01.2022b. 'HC hears PIL on de-notification of Shivalik Elephant Reserve'. *The Times of India*. Retrieved 23.03.2022. https://timesofindia.indiatimes.com/city/dehradun/hc-hears-pil-on-de-notification-of-shivalik-elephant-reserve/articleshow/88718777.cms

06.01.2022c. 'Pune: MoD has cancelled nod to new airport site in Purandar, says Jyotiraditya Scindia'. *The Times of India*. Retrieved 12.11.2022. https://timesofindia.indiatimes.com/city/pune/scindia-mod-has-cancelled-nod-to-new-airport-site-in-purandar/articleshow/88722141.cms

30.01.2022. 'Govt vision for Great Nicobar includes airport & township, some experts think it's nonsense'. *The Print*. Retrieved 20.02.2022. https://theprint.in/environment/govt-vision-for-great-nicobar-includes-airport-township-some-experts-think-its-nonsense/814390/

30.01.2022. 'Land acquisition for airport approach road apace'. *The Hindu*. Retrieved 15.06.2022. www.thehindu.com/news/national/andhra-pradesh/land-acquisition-for-airport-approach-road-apace/article38346841.ece

22.05.2022. 'Jewar's taking off but on ground are broken promises, villages divided and jobless crorepatis'. *The Print*. Retrieved 17.06.2022. https://theprint.in/features/jewars-taking-off-but-on-ground-are-broken-promises-villages-divided-jobless-crorepatis/965603/

25.05.2022. 'Noida Airport seeks suitable partner to create world-class cargo hub near Delhi'. *Business Today*. Retrieved 05.07.2022. www.businesstoday.in/latest/economy/story/jewar-airport-seeks-suitable-partner-to-create-world-class-cargo-hub-near-delhi-334917-2022-05-24

27.08.2023. 'CM reviews status of land acquisition & rehabilitation for Greenfield Airport in Silchar'. *Pratidin*. Retrieved 20.01.2024. www.pratidintime.com/latest-assam-news-breaking-news-assam/silchar/cm-reviews-status-of-land-acquisition-rehabilitation-for-greenfield-airport-in-silchar

Airports Authority of India (2021). Dehradun Airport: New terminal building promo. www.youtube.com/watch?v=VNgr4el3a2U

5 Aerotropolis at what cost, to whom? An analysis of the social and economic impacts of the Yogyakarta International Airport, Indonesia

Ellen Putri Edita

Acronyms

YIA	Yogyakarta International Airport
MP3EI	Masterplan *Percepatan dan Perluasan Pembangunan Ekonomi Indonesia* (Masterplan for Acceleration and Expansion of Indonesia's Economic Development)
RPJMN	*Rencana Pembangunan Jangka Menengah Nasional* (National Medium-Term Development)
ASEAN	Association of Southeast Asian Nations
MPAC	Masterplan of ASEAN Connectivity
PPP	public–private partnership
GDP	gross domestic product
PGA	Pakualaman Ground

Introduction

Under the responsibility of Angkasa Pura, a state-owned company, which has the authority over airport management in Indonesia, the Yogyakarta International Airport (YIA) was built in Temon sub-district, Kulon Progo district, Special Region of Yogyakarta province. The establishment of the YIA became a tangible solution to the prolonged problem that Yogyakarta faced because its old inner-city airport, Adisucipto, could no longer accommodate any more passengers or be expanded (25.06.2015 Liputan 6; Angkasa Pura I 2017; 27.01.2018 Liputan 6). In this regard, the YIA is projected to be able to accommodate up to 14 million passengers per year or around nine times bigger than Adisucipto (Fauzi & Putriani 2017). With bigger capacity, the airport received financial support from GVK Group India through public–private partnership (PPP) scheme and was established on an area of 637 ha that encompassed five villages, which are Jangkaran, Sindutan, Palihan, Glagah, and Kebonrejo (22.12.2016 Detik Finance; Angkasa Pura I 2017). Figure 5.1 shows the ongoing construction of the YIA.

Based on the Masterplan for Acceleration and Expansion of Indonesia's Economic Development (MP3EI), the establishment of the YIA aims to promote Yogyakarta as part of the economic and tourism centres in Indonesia, which can lead

DOI: 10.4324/9781003494966-5

Figure 5.1 The construction of YIA airport.
Source: Photograph by Edita.

to the economic growth of the country. In the bigger picture, the YIA is expected to serve as part of infrastructural development that can facilitate regional connectivity. However, the infrastructural development in Indonesia needs funding from foreign investment, since the country has limited state budget allocated for this purpose (Coordinating Ministry of Economic Affairs 2011).

The establishment of the YIA highlighted different layers of issues. Although the airport has been inaugurated by the president in August 2020 (28.08.2020 *Tempo*), the process behind it was controversial and was therefore postponed for six years (Putri & Paskarina 2021). The idea to transform agricultural land into an airport led to a refusal from the local people to give away their houses and land to the government. Their disagreement was based on the fact that most people who live on the airport land were farmers who were strongly dependent on agricultural land for their livelihood (23.07.2018 *The Jakarta Post*; BPS Statistics of Kulon Progo Regency 2018).

Since the issue of the YIA development came into the surface, land became a promising business entity for the people who owned it or those who had the money to buy it. This is because the price of land on the airport land significantly increased since its inception around 2011/2012 (Utami et al. 2021). In many circumstances, local people who have money started buying land that can be sold later so that they can benefit from multiple sources of profit in the end. When this research was

Social, economic impacts of Yogyakarta International Airport 83

Figure 5.2 Land sales banner along the way to YIA highlighting vicinity to the airport.
Source: Photograph by Edita.

performed in 2019, several areas near the airport had been plotted for building hotels. Furthermore, all the roads that head to the airport were filled with banners that advertised the land. Most of them worked in the same way, highlighting the vicinity of the plots to the airport (Figure 5.2). Land speculation due to the upcoming airport infrastructure raised the real estate prices.

Apart from that, the initial concerns regarding the environmental problems emerged along with the establishment of the YIA. Since 2011, the city of Yogyakarta has experienced a massive growth in the hotel industry. Hotel development required a huge amount of groundwater, which was drawn from the main water sources that were used by the local people in the region (Manifesty 2014). This occurrence led to a groundwater shortage, which was indicated by an annual drop of water level in domestic wells of up to 35 cm (Tifa Foundation 2017). Reflecting on the environmental issues in Yogyakarta in recent years, a similar circumstance could take place in the YIA area and its surroundings because the demand for water would undeniably increase.

The establishment of the YIA was generally brought up in the public discourse because the location of airport land is widely considered vulnerable to natural disasters. High waves and tsunami were the two biggest concerns of the YIA development. The former concern is caused by its proximity to the Indian ocean, while the latter is driven by the airport location, which is situated in the middle of Australian and Euro-Asian tectonic plates. With the disaster-prone attribute of the airport land, environmental activists discouraged infrastructure development because it could exacerbate the current exposure to high waves and tsunami (Chrysanti et al. 2019). The evident risks associated with the airport land, along with the impacts of rapid urbanisation triggered by the airport development, question its rushed implementation.

Given the earlier context, this book chapter analyses the social and economic impacts of the establishment of the YIA. The research question of this study is: What are the social and economic impacts caused by the establishment of the YIA? The analysis addresses the drivers behind the airport establishment, the condition of the region before the land acquisition, the displacement, people's responses, and what they obtained in return.

This research will initially introduce the brief tension that arises from the YIA establishment, which is covered in Section 1. Section 2 provides background information about the national plans and the airport, the political drivers behind the airport establishment, and the livelihood of the people prior to the land acquisition of the airport. Impoverishment risk and reconstruction (IRR), which is employed as the conceptual framework, will be demonstrated in Section 3. Section 4 will provide the findings of the study that encompass the social and economic impacts of the YIA establishment and the political opportunism behind it. Lastly, Section 5 will provide the conclusion.

A qualitative research method was applied to obtain comprehensive insights to reveal the interpretation of the YIA establishment of each relevant participant, since there were different responses towards the decision. In order to collect the primary data, 13 semi-structured interviews, one focus group discussion, and direct observation were conducted. The semi-structured interviews and focus group aimed to obtain perceptions from relevant people who experienced the process of the YIA establishment and to gather different perspectives towards specific cases about the airport. The parties that took part in the interviews and focus group consisted of volunteers in the field, non-governmental organisations (NGOs), researchers, government officials, and local people. Direct observation completed the findings because it brought up small details that were not covered by interview and focus group. The fieldwork focused on noticing land use changes that happened because of the airport construction, the relocation areas for the displaced people, the surrounding natural environment, and small details of the construction area that led to more clues for analysis. Meanwhile, official document review was conducted to attain the secondary data.

Background

The Masterplan for Acceleration and Expansion of Indonesia's Economic Development (MP3EI) and the Yogyakarta International Airport (YIA)

The Masterplan MP3EI was launched in 2011 by the former president of Indonesia, Susilo Bambang Yudhoyono, as a response to regional and global economic competition that creates more challenges for the future development of Indonesia. Located in the centre of new economic gravity, which is East Asia (including Southeast Asia), it would be pertinent for Indonesia to prepare to face this challenge to foster economic development, which could be evenly distributed for all its people. In particular, the MP3EI program aimed to transform Indonesia into one of the developed countries in the world by 2025, which has gross domestic product (GDP) of USD 4–4.5 trillion and economic growth of 7–9% each year (Coordinating Ministry for Economic Affairs 2011).

MP3EI applied a new 'not business as usual' way of thinking, where it highlighted the emphasis of collaboration between all elements in societies and local government (Coordinating Ministry for Economic Affairs 2011; Sahaya & Arto 2012).

Since Indonesia has limited state budget, the accomplishment of economic development of Indonesia is strongly dependent on collaboration with the private sector, including state-owned business and both local and foreign investors (Coordinating Ministry for Economic Affairs 2011). Within MPEI, Indonesia was divided into six main economic corridors: Sumatra, Java, Kalimantan, Sulawesi, Bali-Nusa Tenggara, Papua-Maluku Islands. The development of the economic corridors was performed along with infrastructure development to increase connectivity. As a result, the role of roads, airports, harbours was fundamentally important for achieving the purpose of economic development in Indonesia (Coordinating Ministry for Economic Affairs 2011).

In the end of his leadership in 2014, President Yudhoyono only accomplished projects that were planned in the MP3EI (Indonesia Investments n.d.). Accordingly, after the end of President Yudhoyono's era, the MP3EI's vision was adopted by the new president, Joko Widodo, through National Medium-Term Development Plan (RPJMN) between 2015 and 2019 (08.12.2017 BBC Indonesia). Similar with its predecessor, RPJMN 2015–2019 also developed several strategic regions in the country, which, in the MP3EI era, were called economic corridors. The strategic regions, furthermore, would be developed based on their own potential capacities and connected through acceleration of infrastructure development. In order to enhance connectivity of the strategic regions, RPJMN 2015–2019 targeted to build 2,650 km arterial road, 1,000 km highway, 3,258 km railway, and 15 new airports (Ministry of National Development Planning 2014).

Applying the aerotropolis concept, the government of Indonesia established the YIA, which was initiated as a strategic infrastructure project listed on the MP3EI (Regional Secretary of Special Region of Yogyakarta 2017; Coordinating Ministry of Economic Affairs 2011). An *aerotropolis* itself is defined as an area centred on a major airport and surrounded by the development of non-aviation infrastructures, integrated transportation, and service facilities, which function to gain economic benefit (Kasarda & Lindsey 2012; Kasarda & Appold 2014). Based on this concept, the government of Indonesia planned to develop the area around the YIA as commercial spaces, offices, housings, public facilities, industries, and warehouses (Regional Secretary of Special Region of Yogyakarta Province 2017).

The aerotropolis of the YIA will also be supported by an integrated transportation network, consisting of train access, ring road, and highway that connect the airport with other fundamental transportation facilities (28.08.2021 Kompas; 26.01.2020 Radar Jogja Channel; 27.06.2021 Tempo).

The establishment of the YIA is anticipated to strongly promote the MP3EI's mission in preparing Yogyakarta as part of the economic and tourism centres in Indonesia (Coordinating Ministry of Economic Affairs 2011; 26.08.2017 Kedaulatan Rakyat Jogja). Such a development will encourage a multiplying effect for society since it not only enhances airport service for citizens but also contributes to economic development locally and nationally (Angkasa Pura I 2017). Under the new government, the previously stalled YIA project was embraced and carried forward as part of the RPJMN agenda. In this national plan, Yogyakarta was set as one out of ten of Indonesian's metropolitan cities (Ministry of National Development Planning 2014). President Widodo also included the YIA into one of the national strategic projects since 2016 (Cabinet Secretariat of the Republic of Indonesia 2019), which refer to projects which are regarded as important and urgent for fostering Indonesia's economic growth (Committee for Accelerating Priority Infrastructure Development 2016a).

Besides boosting economic growth, the aerotropolis of the YIA serves as an important role to overcome the existing economic gap in Yogyakarta Province, since it had the highest economic gap in Indonesia for the past years (Regional Secretary of Special Region of Yogyakarta 2018; BPS Statistic of Indonesia 2019). Between four regions within the province, Yogyakarta and Sleman Regency have high regional GDP with 60%, while, in contrast, the GDP in Kulon Progo and Gunung Kidul Regency only reaches 21%. Connectivity, uneven infrastructure development, investment, and low number of tourists are among the key drivers that lead to this massive gap (Regional Secretary of Special Region of Yogyakarta 2018).

The political drivers

The move by the government of Indonesia to boost infrastructure development is closely related to the global economic crisis in 2008 (Basri & Rahardja 2010). Many countries in Asia, including Association of Southeast Asian Nations (ASEAN) countries, were hit by a very sharp fall in their gross domestic product (GDP) in 2009. It was driven by the high dependency of GDP in these countries on exports of goods and services. This collapse, surprisingly, was not experienced by Indonesia, because the export of goods and services was not a big contributor to the national GDP (Rasiah et al. 2014). The advantage of Indonesia in overcoming the 2008 global crisis was caused by the lack of investment and poor quality of logistics that led to slow growth of export. Nevertheless, Indonesia cannot always rely on this good luck, because there is a positive correlation between exports and economic growth, although at the same time this relation can be very risky. Hence, it is fundamentally important for the country to maintain a balance between domestic demand and exports in order to maximise economic growth. Considering that Indonesia is a massive archipelago country, logistics became an essential aspect for Indonesia to connect all parts of the country, including remote areas. Due to poor logistics that drives to unintegrated market, Indonesia was recognised as a country with high cost of transaction, making this country less attractive for investment. Therefore, the national government should prioritise physical infrastructure to reach economic growth from both domestic demand and export sector (Basri & Rahardja 2010).

Besides increasing domestic connectivity, the infrastructure development of Indonesia would also contribute to embody the vision of ASEAN connectivity. Infrastructure development is considered as an important factor in stimulating economic recovery in ASEAN countries after the 2008 global crisis (Pushpanathan 2010). Therefore, in 2010, Masterplan of ASEAN Connectivity (MPAC) was initiated by ASEAN leaders. In principle, MPAC emphasises the promotion of integrated intra-ASEAN trade and the ultimate improvement of connectivity between ASEAN and global market (Pushpanathan 2010; Gunawan 2017).

In Indonesia, the MPAC was integrated in the MP3EI formulation. This means that the economic corridors provided in the masterplan are connected to regional development in ASEAN countries. Therefore, the MP3EI will not only promote national connectivity in Indonesia but also contribute to regional settings through ASEAN connectivity that serves as a bridge to international market (Gunawan 2017).

An integrated economic network, funding, and coordination becomes a key factor to embody MPAC (Zen & Regan 2014). Over the past years, developing countries always relied on government fund for improving infrastructure. However, many economic pressures caused this type of funding to become unstable, leading to less investment for infrastructure. Accordingly, developing countries require other alternatives to fund their infrastructure development (Rillo & Ali 2017). This situation leads ASEAN countries to enhance private sector role by implementing the public–private partnership (PPP) scheme (Zen & Regan 2014). In a more specific case, PPP also takes part in shaping the YIA establishment (22.12.2016 Detik Finance; Committee for Accelerating Priority Infrastructure Development 2016b).

The land and the livelihood

Landownership before the construction of the YIA belonged to several parties. Around 67.23% land was owned privately, 27.3% was Pakualaman Ground (PAG),[1] 5.39% was agency land, and less than 1% was land that was used for other necessities (Utami et al. 2021). Meanwhile, in terms of land use, irrigated rice field and settlement occupied the largest area with 28.96% and 25.80%, respectively. These numbers were followed by swamps and embankment with 17.27% and 11.47%. The land use of the affected villages was closely connected to local livelihood. More than 50% of the local people worked as farmers. Meanwhile, around 24% of the people worked as agricultural labourers, and 5% of the people relied on aquaculture and livestock to make a living. In total, around 80% of the local people in the airport construction area were very dependent on natural resources for their livelihoods (Utami et al. 2021). This means that farmers and people, who relied on natural resources as their source of income, were very vulnerable to the adverse impacts of the massive land acquisition for the airport.

Impoverishment Risk and Reconstruction (IRR)

Many studies across the world have observed severe impacts of airport construction. In North Sumatra, Indonesia, the construction of Kualanamu Airport and its supporting facilities, which required 3,000 ha of productive land in total, caused the farmers loss of livelihoods and culture. Due to the local government's orders to vacate the land, farmers experienced difficulty in accessing essential utilities and public services. They also could not afford to buy land somewhere else and had a similar quality of life because the compensation money they received was very little (Kaputra & Putri 2020).

In a different case, the establishment of Kertajati Airport in West Java, Indonesia, led to the significant reduction of farmers' income. This was rooted in the fact that there was a decrease on farmer's landownership, which influenced the average agricultural production. It was reported that, in total, farmers' income from the agricultural sector was 42.3% lower after the airport's construction (Matufany et al. 2021).

This study adopts the impoverishment risk and reconstruction (IRR) model that was developed by Cernea (2005). The model encompasses the analysis of social and economic impacts from the displacement to the recovery process, where it helps uncover who will benefit and who will suffer from the social and economic impacts of airport development.

The IRR model was adopted in several research that examined the impacts of various project developments, such as by disaster-preventive resettlement engineering in Shaanxi Province, China (Xiao et al. 2018); displacement and resettlement in rutile mining communities in Sierra Leone (Wilson 2019); impoverishment from the post-disaster resettlement project in Cameroon (Bang & Few 2012); or dam-induced displacement in Ethiopia (Eguavoen & Tesfai 2012). The IRR model shows how infrastructure developments, such as industries, irrigation, highways, power generation, and airports, which are accompanied by displacement, may cause eight major impoverishment effects: landlessness, joblessness, homelessness, marginalisation, food insecurity, increased morbidity and mortality, loss of access to common property and services, and social disarticulation. In the context of infrastructure development, the involvement of government actors affects people. This raises the concern about social justice and equity, where some groups of people will gain more benefits than others (Cernea 2005).

Aerotropolis at what cost, to whom?

The social and economic impacts

The establishment of the airport led the local people on the airport land to endure several major detrimental issues. From the IRR model's point of view, the first impact caused by the project was landlessness. The YIA's establishment faced hurdles in land clearing procedure because it entails the process of displacing five villages from the prospective airport land. The displacement, which came first in the overall YIA establishment, was probably the most controversial step of the project because it highlighted the struggle of local people to defend their land (08.12.2017 BBC Indonesia; 23.07.2018 The Jakarta Post; 25.07.2018 CNN Indonesia). This process divided the local people into two groups: people who agreed with the displacement and people who disagreed. This led to the boycotting of people who disagreed to be displaced. During the interview, an NGO indicated that there was misinformation in introducing the YIA to local people. Local people who opposed to be displaced were not allowed to join discussions between Angkasa Pura and the population. They were not allowed to enter the discussion room even though they brought the official invitation letter from Angkasa Pura.

In contrast, another version comes from a village officer, who perceived that people who opposed displacement always refused to come to meet with Angkasa Pura, although they had been invited many times. Responding to this, another interviewee argued that, in practice, coming to the meetings with Angkasa Pura was sometimes considered as an agreement towards the airport. This finding probably became the reason that people who opposed displacement refused to join the discussion of airport establishment. People knew that the YIA played a crucial role in the national development and that the project would still continue irrespective of their participation in public meetings with Angkasa Pura. Besides the issue of exclusion, the local people also experienced oppression and violence. Some interviewees said:

'The people were dragged to leave their houses.'
'They use military practice to force them to leave.'
'They even ruined my agricultural yield that had just been harvested.'

On the other hand, Angkasa Pura had a tight deadline to clear the airport land because President Widodo, through a law, had officially instructed the company to accelerate construction and operation of the megaproject in order to finalise the airport until April 2019 (GovIndonesia 2017). Obtaining much support from relevant stakeholders, even from the president himself, Angkasa Pura successfully emptied the airport land from settlements. This means that the local people had no other option but to leave.

People were relocated in the areas that had been identified by Angkasa Pura. In the case of the YIA, one thing that needs to be underlined is that just because people agreed to be relocated did not necessarily mean that they agreed with the airport establishment. Instead, they did that for the sake of the country. One of the local people said:

> *'We are here now not because we agreed with the airport. Since the airport is built for the sake of the country, we have no option. We sacrificed our land. Therefore, our land is not sold for us. It is needed by this country.'*

He also mentioned that he still needed to pay the land as well as construction cost of his new home in the relocation area. Although some people think that their existing new houses and neighbourhood were more organised (Figure 5.3), he thought that some elements were irreplaceable in comparison with his previous life, referring to his urban neighbourhood, which cannot be cultivated.

In terms of compensation, there was a treatment gap experienced by the local people. During the land clearing process, the land and property of the affected people were appraised by Angkasa Pura to determine how much money they would pay as compensation. It was found that the appraisal process strongly depended on people's reaction towards relocation. People who agreed to be relocated received the agreed compensation. In contrast, people who refused only received a little amount in the end. The meagre money was not enough to support their families. Transparency was another issue that differentiated the appraisal experience. One man reported that his house, yard, and agricultural land were counted separately from each other during the appraisal. The process seemed fair, until he questioned about the price given by Angkasa Pura for each asset. He found that the price

Figure 5.3 Relocation area of Palihan.
Source: Photograph by Edita.

differed significantly from one house to another, although previously they were located in the same neighbourhood.

The second impact that needed to be endured by the local people was joblessness. When this research was conducted, people who agreed to be displaced were still waiting for the responsibility of Angkasa Pura to give them new jobs, as was promised previously. Before the displacement process, people from Angkasa Pura had mentioned that local people would be prioritised to receive jobs from the airport. The company also performed surveys with the local people to collect data about what kind of work skills they had and what skills they wanted to develop in the future. By having that information, Angkasa Pura promised to facilitate training centres that could be used to prepare local people to work in the airport during construction or operation. In less than a year, after the people were displaced and the construction of the airport began (Putri & Paskarina 2021), Angkasa Pura said that only people below 35 years could be included in training centres. This was very unexpected, considering that many farmers were older. The surveys about work skills did not change anything, because until the interview was conducted, there was no work call from Angkasa Pura.

The study about the economic impacts from the YIA establishment shows that after the land acquisition, 15% of the respondents remained unemployed. This group of people previously worked as agricultural labourers, tenant farmers, breeders, and embankment workers. Meanwhile, 9% of the respondents did not have permanent jobs. About 7% of people eventually took part in the airport construction; their future employment, however, was unclear. From the perspective of income security, some people still had a lot of money from the compensation. However, without good financial planning and income sources, this amount of money would not last long. Due to the existence of the new airport, the living cost in the affected villages increased. The main factor was higher electricity cost. Unlike before the airport was established, people could no longer cultivate their own land to fulfil their food needs. They need to buy everything to fulfil their daily needs (Utami et al. 2021).

The situation had aggravated when the Corona-19 pandemic hit Indonesia in 2020. Displaced people struggled to earn money. The small businesses they developed ran out of customers due to the pandemic. At the same time, they had spent all the compensation to buy land and cars. Many children dropped out of schools because their parents did not have enough money to afford their education (02.05.2020 Tempo). This proves that due to the airport development, some people experienced marginalisation. They no longer have access to their own land, the promised training opportunities, or school education to ensure their living.

The concern about sustainable livelihoods was closely related with the loss of identity that farmers faced. One of the interviewees who opposed to be displaced shared how being a farmer meant everything for him. He mentioned:

'As farmers, we can manage our own working time. We are very flexible because we do not work for other people.'

Another interviewee added:

'Farmers is a reliable job. By far, this job can support my family. Having our own land to be planted is enough for us.'

People who opposed the displacement put a high value on their land because it is associated with their identity as farmers, which enabled them to work independently and support their families. The value of farmland and the loss of identity as farmers were not acknowledged by Angkasa Pura and other YIA stakeholders. Angkasa Pura perceived that land acquisition and interaction with farmers were over after having paid the compensations. But the impacts of displacement were felt in different domains as well.

From the perspective of the IRR model, it was observed that people who were displaced for the airport construction suffered from landlessness, joblessness, and marginalisation. Other impacts, including homelessness, food insecurity, increased morbidity and mortality, loss of access to common property and services, and social disarticulation were not found in this case.

The political opportunism

The YIA was officially inaugurated by President Joko Widodo in the end of August 2020. On that occasion, the president praised the fast process of the overall airport establishment, which only required 20 months (28.08.2020 Liputan 6). It is interesting to see that the rapid establishment of the YIA was made possible by President Joko Widodo. He seems to have high concern in this national strategic project by generating Law Number 98 of 2017, which specifically aims to accelerate construction and operation of the YIA megaproject. From there, the instruction to finalise YIA in April 2019 is affirmed. Although the timeline was behind the expected target, the project concluded swiftly since then. Looking at the law in more detail, why did the president choose April 2019 as the due date?

The impacts of the YIA establishment in both short- and long-term periods should have come under meticulous government consideration since this project aimed to bring compounding economic growth for Yogyakarta and Indonesia through aerotropolis development. In practice, the rapid construction of the YIA shows that this airport should be finished as soon as possible. At the same time, every five years, Indonesia holds general elections for president, vice president, and legislative members. President Widodo, who had served as the leader of this country for the past period, nominated himself again in the general election, which was held in April 2019. This proves that the YIA was used by him as a political strategy for the following elections. Taking over the project that was always delayed in the previous leadership and creating a sudden breakthrough through Law Number 98 of 2017 were the tangible efforts by President Widodo to raise up the popularity of the YIA in the public. In the end, his strategy to highlight the airport establishment turned out to be working. Joko Widodo was re-elected as the president of Indonesia from the national election in April 2019. Possibly, among the huge number of his voters, there were people who admire his performance towards the YIA.

Conclusion

Since the launch of the MP3EI, which aimed to foster economic development of Indonesia, the YIA has become a very important agenda for the president. The airport was initially designed to prepare Yogyakarta to be the centre of economic and tourism so it can be connected to a bigger market and accommodate more tourists in the future. Like other infrastructure development projects, the construction of the YIA was not without controversy. The study found that the massive land acquisition to build the airport left the affected villages encountering severe social and economic impacts. The complexity of the YIA construction was analysed with the IRR model, which focuses on the impoverishment effect that happens from the forced displacement caused by infrastructure development.

Referring to the IRR model, there are three main impacts that affected people who lived on the airport land before construction of YIA: landlessness, joblessness, and marginalisation. Many people lost their agricultural land and settlement because of the airport construction. Furthermore, the compensation money received by the people was not equitably distributed so that it created rifts within the communities. At the same time, the people could not make a living as farmers because the amount of farmland in the relocation site is very limited, and its price is increasing. At this moment, only certain households can afford to buy land near the airport. Getting jobs in the airport is almost impossible because it requires specific age and education level, which are not owned by most of the displaced people. Statistics show that the economic development in Kulon Progo (the district where the YIA is located) significantly increased after airport construction. In 2018, the economic development in this region reached 11.3% in 2018. Prior to 2018, the economic development in Kulon Progo was only around 4–5.2% (30.08.2020 Tempo). This explains that the airport, initially built to help economic growth in Indonesia, has already created local inequality. The IRR model helps to understand that while local incomes show significant growth in general, some group of people will remain poor because they are left without any choice.

Besides enduring the risk of impoverishment, the local people also experienced violence during the displacement process and loss of identity as farmers. The latter is not acknowledged by the relevant YIA stakeholders because they believe that providing a compensation to former residents of the airport land was sufficient to compensate their loss of land. Lastly, the performance of the president of Indonesia, who integrated YIA to the current national plan and national strategic projects and mandated a special law to accelerate airport construction, indicates that the president expected the political return for the next election agenda in 2019, which he won indeed.

Note

1 The PAG refers to the land that is managed and cultivated by Kadipaten Pakualaman (the name of the dominant political system in the region before the independence of the country) and its people (Fitriantoro 2020).

References

Angkasa Pura I (2017). *Analisis dampak lingkungan hidup (ANDAL) pembangunan bandar udara Yogyakarta International Airport* [Environmental impact analysis (ANDAL) for the construction of the Yogyakarta International Airport]. Unpublished company report. Jakarta: PT Angkasa Pura I (Persero).

Bang, H. N. and R. Few (2012). 'Social risks and challenges in post-disaster resettlement: The case of lake Nyos, Cameroon'. *Journal of Risk Research* 15(9), 1141–1157. https://doi.org/10.1080/13669877.2012.705315

Basri, M. C. and S. Rahardja (2010). 'The Indonesian economy amidst the global crisis: Good policy and good luck'. *ASEAN Economic Bulletin* 27(1), 77–97.

BPS Statistic of Indonesia (2019). *Gini ratio provinsi 2002–2018* [Provincial gini ratio 2002–2018].

BPS Statistics of Kulon Progo Regency (2018). *Temon subdistrict in figures*.

Cabinet Secretariat of the Republic of Indonesia (2019). Pembangunan sesuai target, bandara 'New' Yogyakarta akan beroperasi mulai April. [Reaching construction target, the 'New' Yogyakarta airport will start operating in April]. Retrieved 19.02.2024. https://setkab.go.id/pembangunan-sesuai-target-bandara-new-yogyakarta-akan-beroperasi-mulai-april/

Cernea, M. M. (2005). 'Concept and method. Applying the IRR model in Africa to resettlement and poverty'. In *Displacement risks in Africa. Refugees, resettlers and their host population*. I. Ohta and G. Yntiso (eds.). Kyoto: Kyoto University Press, 195–258.

Chrysanti, A., M. B. Adityawan, B. P. Yakti, J. Nugroho, K. Zain, I. Haryanto, M. Sulaiman, A. Kurniawan, and H. Tanaka (2019). 'Prediction of shoreline change using a numerical model: Case of the Kulon Progo Coast, Central Java'. *MATEC Web of Conferences* 270, 04023.

Committee for Accelerating Priority Infrastructure Development (2016a). *Proyek strategis nasional* [National strategic project]. Jakarta.

Committee for Accelerating Priority Infrastructure Development (2016b). *Pembangunan baru bandara internasional di propinsi D.I. Yogyakarta* [Construction of a new international airport in the special region of Yogyakarta province]. Jakarta.

Coordinating Ministry for Economic Affairs (2011). *Masterplan for acceleration and expansion of Indonesia economic development 2011–2025*. Jakarta.

Eguavoen, I. and W. Tesfai (2012). 'Social impact and impoverishment risks of the Koga irrigation scheme, Blue Nile basin, Ethiopia'. *Afrika Focus* 25(1), 39–60.

Fauzi, I. and D. O. Putriani (2017). *Kajian transportasi umum pra dan pasca pengoperasian Yogyakarta International Airport (YIA)* [Pre-and post-operation public transportation studies of Yogyakarta International Airport (YIA)].

Fitriantoro, M. I. (2020). 'Drivers of conflict in urban infrastructure: Case study of the new Yogyakarta Airport'. *Jurnal Politik* 6(1), 89–122. https://doi.org/10.7454/jp.v6i1.214

GovIndonesia (2017). Peraturan Presiden Republik Indonesia Nomor 98 Tahun 2017 tentang percepatan pembangunan dan pengoperasioan bandar udara baru di Kabupaten Kulonprogo Provinsi Daerah Istimewa Yogyakarta [Presidential Regulation of the Republic of Indonesia Number 98 of 2017 concerning the acceleration of the construction and operation of new airports in Kulonprogo Regency, Special Region of Yogyakarta]. https://setkab.go.id/wp-content/uploads/2017/10/Perpres-98-Tahun-2017.pdf

Gunawan, G. H. (2017). *A new dynamism of regionalism towards ASEAN connectivity and its linkage with national and local development projects: A case study of Indonesia*. Doctoral thesis. International Public Policy, Tsukuba: University of Tsukuba.

Indonesia Investments (n.d.). Masterplan for acceleration and expansion of Indonesia's economic development. www.indonesia-investments.com/projects/government-development-plans/masterplan-for-acceleration-and-expansion-of-indonesias-economic-development-mp3ei/item306

Kaputra, I. and P. W. Putri (2020). 'The precarity of peri-urban resistance: A resistance to the forced eviction of Pasar VI village and the development of Kualanamu International Airport, North Sumatera'. *PCD Journal* 8(1), 49–67.

Kasarda, J. D. and S. J. Appold (2014). 'The economics of international airline transport', p. iii. https://doi.org/10.1108/S2212-160920140000004015

Kasarda, J. D. and G. Lindsey (2012). *Aerotropolis. The way we will live next*. New York: MacMillan.

Manifesty, O. R. (2014). 'Rapid growth of hotels in Yogyakarta and its relation to the city's water supply'. *e-Journal of Tourism* 6(1), 43–51. https://doi.org/10.24922/eot.v6i1.41878

Matufany, O. M., Istiqomah, N. Kadarwati, and S. D. Purnomo (2021). 'The impact of airport construction on farm households' income'. *Journal Ekonomi & Studi Pembangunan* 22(1), 1–11.

Ministry of National Development Planning (2014). *Rencana pembangunan jangka menengah nasional 2015–2019* [National medium term development 2015–2019]. Jakarta.

Pushpanathan, S. (2010). ASEAN connectivity and the ASEAN economic community. June 7.

Putri, D. N. C. and C. Paskarina (2021). 'Strategies to de-escalate the conflict of Yogyakarta International Airport construction'. *MIMBAR Journal Social dan Pembangunan* 37(1), 57–66.

Rasiah, R., K. C. Cheong, and R. Doner (2014). 'Southeast Asia and the Asian and global financial crises'. *Journal of Contemporary Asia* 44(4), 572–580. https://doi.org/10.1080/00472336.2014.933062

Regional Secretary of Special Region of Yogyakarta Province (2017). Yogyakarta International Airport (YIA): Mewujudkan DIY sebagai daerah tujuan wisata dan budaya kelas dunia. [Yogyakarta International Airport (YIA): Creating Special Region of Yogyakarta province as a tourist destination and world-class culture]. Retrieved 19.02.2024. http://bappeda.jogjaprov.go.id/download/download/379

Regional Secretary of Special Region of Yogyakarta Province (2018). *Ketimpangan wilayah DIY*. [Inequality of special region of Yogyakarta], Presentation.

Rillo, A. D. and Z. Ali (2017). *Public financing of infrastructure in Asia: In search of new solutions*. ADI Policy Brief 2017–2. Tokyo: Asian Development Bank.

Sahaya, H. N. and A. Arto (2012). 'Peran MP3EI berbasis "not business as usual" untuk meningkatkan daya saing dan penyerapan tenaga kerja' [The role of MP3EI is based on "not business as usual" to improve competitiveness and employment]. *Jejak* 5(1), 46–59.

Utami, W., D. Nurcahyanto, and S. Sudibyanung (2021). 'Economic impacts of land acquisition for Yogyakarta International Airport project'. *MIMBAR Journal Social dan Pembangunan* 37(1), 150–160. https://doi.org/10.29313/mimbar.v37i1.6955

Wilson, S. A. (2019). 'Mining-induced displacement and resettlement: The case of rutile mining communities in Sierra Leone'. *Journal of Sustainable Mining* 18(2), 67–76. https://doi.org/10.46873/2300-3960.1087

Xiao, Q., H. Liu, and M. Feldman (2018). 'Assessing livelihood reconstruction in resettlement program for disaster prevention at Baihe County of China: Extension of the impoverishment risks and reconstruction (IRR) model'. *Sustainability* 10(8), 2913. https://doi.org/10.3390/su10082913

Zen, F. and M. Regan (2014). *Financing ASEAN connectivity* ERIA Economic Research Institute for ASEAN and East Asia: Jakarta.

Media sources

25.06.2015. 'Alasan Yogyakarta harus miliki bandara baru' [The reason why Yohyakarta must have a new airport]. *Liputan 6. Alasan Yogyakarta Harus Miliki Bandara Baru – News Liputan6.com* (retrieved on 16.02.2024).

22.12.2016. 'Investor India akan garap badara Kulon progo di Yogyakarta' [Indian investor will work in the Kulon Progo airport in Yogyakarta]. *Detik Finance. Investor India akan Garap Bandara Kulon Progo di Yogyakarta (detik.com)* (retrieved on 16.02.2024).

26.08.2017. 'Bandara YIA, jadi pintu gerbang utama pariwisata RI' [YIA Airport, becomes the main gateway for Republic of Indonesia's tourism]. *Kedaulatan Rakyat Jogja*.

08.12.2017. 'Bandara Kulon progo, Yogyakarta: antara sabda leluhur dan "proyek strategis" Jokowi' [Kulon Progo Airport, Yogyakarta: between the words of the ancestors and Jokowi's "strategic project"]. *BBC Indonesia*. *Bandara Kulon Progo, Yogyakarta: antara sabda leluhur dan 'proyek strategis' Jokowi – BBC News Indonesia* (retrieved on 16.02.2024).

27.01.2018. 'Kapasitas terbatas, Bandara Adisutjipto makin padat dan sesak' [Capacity is limited, Adisutjipto Airport is getting more crowded]. *Liputan 6*. *Kapasitas Terbatas, Bandara Adisutjipto Makin Padat dan Sesak – News Liputan6.com* (retrieved on 16.02.2024).

23.07.2018. 'Last 86 families still fighting against Yogyakarta airport development to keep land. *Jakarta Post*. *Last 86 families still fighting against Yogyakarta airport development to keep land. National – The Jakarta Post* (retrieved on 16.02.2024).

25.07.2018. 'Warga Kulon Progo tatang buldoser lawan penggusuran' [Kulon residents challenge bulldozer against displacement]. *CNN Indonesia*. *Warga Kulon Progo Tantang Buldoser Lawan Penggusuran (cnnindonesia.com)* (retrieved on 16.02.2024).

26.01.2020. 'Resmi dibuka, underpass bandara YIA jadi kebanggaan warga Kulonprogo' [Officially opened, YIA underpass became the pride of the Kulonprodon's residents]. *Radar Jogja Channel*. Video. *Resmi Dibuka, Underpass Bandara YIA Jadi Kebanggan Warga Kulonprogo – YouTube* (retrieved on 16.02.2024).

02.05.2020. 'Pontang-panting setelah tergusur Bandara Kulon Progo' [Helter-skelter after being displaced by Kulon Progo Airport]. *Tempo*. *Pontang-Panting Setelah Tergusur Bandara Kulon Progo – Nasional Tempo.co* (retrieved on 16.02.2024).

28.08.2020. 'Jokiwi piji pengerjaan bandara Kilon Progo, terbaik di Indonesia' [Jokowi praises the construction of Kulon Progo Airport, the best in Indonesia]. *Tempo*. *Jokowi Puji Pengerjaan Bandara Kulon Progo, Terbaik di Indonesia – Bisnis Tempo.co* (retrieved on 16.02.2024).

28.08.2020. 'Resmikan badara baru Yogyakarta, Jokowi apresiasi pembangunan yang super cepat' [Inaugurating Yogyakarta, Jokowi appreciates it super fast development]. *Liputan 6*. *Resmikan Bandara Baru Yogyakarta, Jokowi Apresiasi Pembangunan yang Super Cepat – Bisnis Liputan6.com* (retrieved on 16.02.2024).

30.08.2020. 'Bandara YIA dibangun, pertumbuhan ekonomi Kulon Progo Melonjak jadi 11.3 persen' [YIA was built, economic growth in Kulon Progo soars to 11.3 percent]. *Tempo*. *Bandara YIA Dibangun, Pertumbuhan Ekonomi Kulon Progo Melonjak jadi 11,3 Persen – Bisnis Tempo.co* (retrieved on 16.02.2024).

27.06.2021. 'Jalan tol Solo Yogya Bawen yang hubungkan 3 bandara bakal beroperasi 2024' [Solo Yogya Bawen highway that connects 3 airports will operate in 2024]. *Tempo*. *Jalan Tol Solo Yogya Bawen yang Hubungkan 3 Bandara Bakal Beroperasi 2024 – Bisnis Tempo.co* (retrieved on 16.02.2024).

28.08.2021. 'Fakta seputar KA bandara YIA' [Facts about YIA train]. *Kompas*. *Fakta Seputar KA Bandara YIA: Harga Tiket, Rute, dan Jadwal Operasi Halaman all – Kompas.com* (retrieved on 16.02.2024).

Tifa Foundation (2017). 'Jogja darurat air' [Water emergency in Jogja] [Video file]. March 22. www.youtube.com/watch?v=WyKUVydtvZc&t=704s

6 The popular appropriation of the airport reserve in Abidjan, Côte d'Ivoire, and strategies to resist displacement

Irit Ittner

Introduction

In January 2020, security officers at Félix-Houphouët-Boigny (FHB) International Airport did not notice, and therefore could not impede, a 14-year-old boy entering the airport grounds in order to hide himself in the landing gear of a plane going to Paris. Ivorian and French news mourned his fatal journey and debated the unlikeliness of this occurring. Many people in Abidjan were shocked by the event while facing conflicting opinions to the political responses that followed. The Ivorian Ministry of Transport, determined to prevent such incidents in the future, gave the order to the private airport operation enterprise AÉRIA to enforce the security zone. There was desperation and resignation, as well as lamentation and rage, when residents in Adjouffou had to leave their homes for a 200 m wide strip to be bulldozed parallel to the wall surrounding the airport grounds. While Adjouffou residents and some journalists subscribed to the view that this eviction was unjust and a threat to human rights, many other *Abidjanais* were in complete agreement with the eviction. In their opinion, Adjouffou residents had illegitimately been there. They welcomed the clampdown against illegal squatting, at times with anti-poor or xenophobic undertones. Later in this chapter, we will return to the residents of the security zone. The eviction from their homes, however, is part of a long history of urban displacement that accompanied the evolution of Abidjan. The metropolis was founded as a French colonial capital in 1903. It served as the capital of the independent Côte d'Ivoire until 1983. Today, Abidjan is a coastal metropolis with a mixed West African, European, and Lebanese population of 5.61 million inhabitants and very high economic ambitions, as well as vast social-economic inequalities.

This book chapter tells how the Ivorian government is strengthening the FHB airport in order to boost the national economy through tourism and a revived aviation sector. It also narrates how various local interest groups appropriated the demarcated land for airport expansion (referred to as the airport reserve) since the 1950s and how co-evolution between the airport and the city resulted in two distinct spatial forms: peri-urban agriculture and dense urban housing. The study also explains the eviction of Adjouffou residents in the larger framework of the urbanisation of Port Bouët, the respective municipal district of Abidjan.

DOI: 10.4324/9781003494966-6

The chapter is guided by the following questions: How did the appropriation of land resources around the FHB International Airport by diverse actors lead to the contemporary spatial forms on the airport reserve? And with regard to Adjouffou, what strategies did the displaced residents use in order to negotiate the persistence of their settlement within the security zone of the airport?

Primary data consisted of observations noted in research diaries, photographs taken during photo walks, and semi-structured interviews conducted during six intense weeks of ethnographic field research in November and December 2021. Analysis also draws from Ivorian newspaper articles and historical photographs. The description of historical land appropriation is based on earlier studies conducted by Ivorian scholars and the author.

The chapter continues with a background section on Ivorian aviation and the central role played by the FHB airport in development visions. This is followed by a conceptual section. The following sections present agriculture and unplanned housing in a historical perspective and describe resistance strategies in Adjouffou.

The chapter concludes that ongoing contestations on airport land[1] in Abidjan originated from restrictions to Ivorian citizenship and the barring of foreigners[2] from legal landownership, as well as from the illegal appropriation of the airport reserve. *Laissez-faire* politics and eviction policy by public authorities also contributed to contestations. Practices of urban citizenship turned the airport reserve into an intensely used urban and peri-urban landscape, with dynamic politics and high competition over land resources. This situation on the ground is hindering the government's attempts to implement the airport city.

The FHB International Airport

Côte d'Ivoire, a West African country with more than 29 million inhabitants, has 27 military and civil airports/airfields, including three international airports. Six civil airports underwent modernisation under a government programme from 2014, including the airport in San Pedro, where an aviation-related eviction of more than 260 households was reported in April 2019.[3] Evictions and resettlement of villages were also scientifically documented in the case of airport expansion in Bouaké (Sanogo & Doumbia 2021).

The largest airport covered by the government's modernisation programme, the FHB International Airport, is situated in Port Bouët Municipality, about 6 m above sea level between the Ebrié Lagoon and the Atlantic Ocean (see Figure 6.1).

The municipality's land is flat and swampy, with inland valleys. Urban farms, gardens, watersides of the lagoon, and beaches are characteristic, as well as large areas covered by factories and warehouses, in addition to French military zones. Port Bouët Municipality had about 618,792 inhabitants in 2021.[4] Many residents are industrial workers, students, farmers, fishermen, or people employed in the informal sector who live on very low budgets. Other residents work around the airport or trade across West African transnational networks. In 2013, about one-third of housing in Port Bouët was situated in precarious quarters. Two-thirds of the residents were tenants (Fall & Coulibaly 2016: 52). Farmland, waterfronts,

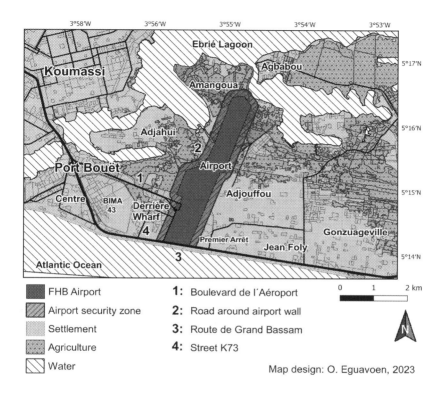

Figure 6.1 Location of the FHB International Airport.
Source: Ittner (2023).

and the proximity to the airport attracted the attention of the national government, estate and tourism developers, as well as business consultants. Though their specific objectives differed, they formed a coalition that envisages Abidjan regaining its pre-conflict importance as the economic hub of West Africa with international business tourism.[5] The coalition perceives the expansion of aviation and the construction of an airport city as crucial ingredients to realising this target.

While the history of aviation in Côte d'Ivoire began in the 1920s based on the use of the Ebrié Lagoon as a marine aerodrome, the history of the terrestrial FHB airport began in 1939, when Côte d'Ivoire was part of colonial French West Africa (Reynaud-Matheis 1962: 79). Kouassi (2021) argued that the Ivorian aviation sector and the FHB underwent six periods characterised by specific historical conditions and developments.[6] In 2022, during the ongoing 'period of revival', FHB airport was one of the fastest-growing African airports in terms of passenger numbers and cargo. In 2017 and 2019, it served more than 2 million air travellers. The government under President Alassane Ouattara, who began his third term in 2020, envisioned an increase to 8 million air travellers to and from Abidjan.[7] Supportive policies were implemented, such as the winding up of the insolvent national airline, Air Ivoire, and the foundation of the novel national airline Air Côte d'Ivoire in 2012. Abidjan was repositioned as the central platform of a West and Central African hub-and-spoke network (Kouassi 2021: 79f.).[8] Passenger numbers of FHB were on the rise, until the Covid-19 pandemic reduced international air travel in 2020.

But how was the land for the airport appropriated at first? In 1939, when terrestrial aviation began in Port Bouët, the local economy was dominated by a wharf. A few buildings dominated the airport grounds, including the aerodrome and the runway. On historical aerial photographs from the 1950s, the land beyond the airport grounds seems to have been covered by natural vegetation (Mairie d´Abidjan 1963: 37). Only a few people lived around the vicinity of the airport, including employees of the airport. In 1979, the Ivorian government, under President Félix Houphouët-Boigny, formally designated a large area in Port Bouët to the airport, including a large land reserve for future expansion.[9] The airport decree from 1979 was renewed in 2010 and reaffirmed these lands to be state land.

In line with the political era of state capitalism in Côte d'Ivoire after independence, the airport was first managed as a public enterprise. When the Ivorian economic miracle began to collapse in the mid-1980s and President Félix Houphouët-Boigny died in 1993, his successor, President Henri Konan Bedié, coming to power in 1995 and facing empty treasuries, took a different political course. In 1996, when airports in Côte d'Ivoire were privatised, the government gave out a 15-year-long management concession to Egis, a French multinational entrepreneurial group with a broad portfolio, which handed the operation of the airport to AÉRIA, an Ivorian company and part of the Egis group. The land concession for the airport land runs until 2030. The airport ground and land reserve have become the quasi-private property of Egis for the duration of the concession. In everyday parlance, most people living and working around the airport told us that the state owned the land, without paying too much attention to the concession.

The international airport of Abidjan is mainly owned by private entrepreneurs and investors, with a minor public share.[10] However, the airport began to exceed the airport grounds as various large actors and interest groups established a mosaic of aviation-related uses,[11] in addition to French military zones and other land uses on the reserve. Non-aviation-related uses included urban agriculture, housing, and manifold commercial activities. Before discussing agriculture and unplanned housing in detail and returning to Adjouffou, we will need to introduce the basic conditions of land tenure and belonging in Abidjan.

The governance of land and claims of belonging in Côte d'Ivoire

Land governance

Land tenure in Abidjan is pluralistic. It draws from state law and customary rules of the Ebrié, which overlap and, at times, conflict with each other. In addition, there is a local regulatory body going back to none of these sources. West African migrants introduced it. They institutionalised new rules for land transaction through repeated practices of exchanging, inheriting, renting out, and selling land. Their leaders, referred to as neo-customary leaders, have been described as actors of numerous social-legal transformations in the city and who are politically powerful (Shighata 2017: 13f, 39; Chéke et al. 2018). Their performance of land governance is contested by the state and the Ebrié but supported by the municipal council of Port Bouët (see later text). Legal pluralism allows forum shopping; individuals choose the body of rules that best suits their interests. They also tend to politically ally with authorities who govern these bodies of rules.

While land legislation in Côte d'Ivoire was reformed in 2013 with the introduction of the *Arrêté de Concession Définitive* (ACD), which is the only legally accepted and protected private landownership title since, many land transactions in Abidjan continue to follow the customary rules of the Ebrié. This population settled around the lagoon long before French colonisation and the foundation of Abidjan. Their customary territories consist of large areas around the 27 villages in the city and extend to water spaces. Villages are administered by *chefferies* (chieftaincy councils). Legally, Ebrié villages and customary territories form rural enclaves today, which the metropolis has spatially overgrown. Land transactions in Ebrié territories combine collective with individual bundles of rights. Their practice and social acknowledgement are determined by the social relations, embedded networks, and status of rights-holders within local Ebrié communities.[12]

In Abidjan, as in rural Côte d'Ivoire, the *tutorat* is a typical institution for long-term land transaction between first-comers (autochtons) and latecomers (migrants). Latecomers receive land rights after entering an enduring patron–client relation with the natives, who act as tutors. Land rights stemming from a *tutorat* can also be inherited. For land-takers, *tutorat* contracts entail a duty of gratefulness, social integration, and political followership of local authorities. Even if land transactions are monetarist and documented by receipts, customary land transactions remain socially embedded land transactions. Tutors continue to own the land. In Port Bouët, it was explained that following Ebrié generations could not use the customary land any longer due to enduring *tutorat* contracts. This was one source of conflict between younger Ebrié and land-takers.

The Ivorian land law from 1998 aims 'to identify, recognize and protect rights acquired through customary transfers but excludes foreigners from land ownership' (Chauveau & Colin 2010: 86). The land law thus strengthened the administrative rights of land users and opened an option for the registration of landownership of land-takers, to the disadvantage of tutors. At the same time, the law prevented landownership for non-Ivoirians.

People excluded from legal access to building plots in Port Bouët find appropriate plots via squatting, customary or neo-customary transactions. These transactions are widely practised. In Port Bouët, some West African migrant communities received land from Ebrié tutors in order to establish farms or *campements* (camps).[13] Some of these settlements grew massively in population, turned into unplanned settlements, and were formally recognised by the municipal council as urban quarters of Port Bouët. Adjouffou became one such urban quarter. Other settlements received the legal status of an urban village. Leaders of urban villages and urban quarters act autonomously and cooperate with municipal authorities. In urban villages, they govern land issues, such as the issuing of temporal use rights (*attestation villagoise*), and land administration. In the context of airport expansion, an increase in land conflicts between tutors and former *campements* was observed (Ittner 2023).

The Ivorian civil war was, among other factors, triggered by related questions. Heirs of tutors questioned the legitimacy of long-term *tutorat* contracts, feeling that these hindered them from accessing their ancestors' lands. Scholars interpreted the Ivorian civil war as a 'war of modernity', which contested the definition of Ivorian belonging and the access to legal rights attached to the Ivorian passport, such as political participation and landownership (Baégas & Marshall-Fratani 2007: 83). Landownership and belonging, therefore, remain very sensitive political issues in the country (Bredeloup 2003; Marshall-Fratani 2006; Zina 2017).

In Abidjan, the temporal allocation of customary land use rights, the so-called *location provisoire* (temporal rental), has gained prominence. Rents generate income for Ebrié landowners but, perhaps more importantly, exercise leadership over customary territories. Obviously, territorial boundaries of Ebrié overlap neither with municipal borders nor with the designated limitation of the airport land. The Ebrié village of Ancien Koumassi, for example, spreads over areas of Marcory and Port Bouët, including minor parts of the airport reserve. The Ebrié village of Agbabou in Port Bouët stretches over a major part of the airport land, including the airport grounds. Customary landowners initially had to be included in decision-making over the airport. When the airport land became a public utility in 1979, the government compensated the Ebrié villages for the expropriation of customary land. The municipal council in Port Bouët legally holds no decision-making power over the public utility. By formally acknowledging unplanned settlements located on the airport reserve and the integration of their leaders into the municipal council, the municipal government practically accessed and governs four populous urban quarters located east of the airport grounds: Adjouffou, Derrière Wharf, Jean Foly, and Gonzuageville.

Claims of belonging

Citizenship is perceived as a political, legal, social, and cultural institution (Isin 2009: 371). In this chapter, I consider citizenship as a legal status and everyday practice in order to better understand the land use dynamics on the airport reserve. Urban citizenship is conceptualised independently from the national citizenship of urbanites. It refers to the political claim-making and the everyday exercising of rights, such as the rights to housing, work, mobility, or political participation. By occupying and inhabiting urban spaces (from the perspective of the authorities in an illegal or anarchic way), residents practice urban citizenship.

The understanding of citizenship in the legal sense is also important for the Ivorian case study because of the restrictive land law from 1998. According to the census from 2021, 22% of the resident population in the country did not hold

Ivorian citizenship. Apart from many Ivorians and *Abidjanais* with foreign, mainly West African, passports, Côte d'Ivoire hosts the largest population of stateless people worldwide. The United Nations Refugee Agency estimates the number at more than 955,000 people (Rosa-Luxemburg-Stiftung 2020). The French enforced and encouraged labour migration from other African French colonies to Côte d'Ivoire. After independence in 1960, immigration continued. Statelessness affects mainly the children and grandchildren of West African migrants because of a very rigid nationalisation law. Many people also lost their documents when fleeing to Abidjan during the civil war because they felt safer in the city (Rosa-Luxemburg-Stiftung 2020).

Agriculture on swampy sections of the airport reserve

Most of the western and northern parts adjacent to the airport wall remain open peri-urban spaces with rural settlements (see Figure 6.1). The local economy is dominated by agriculture, poultry, and rabbit rearing, as well as by artisanal fishery. Many residents commute to other parts of Port Bouët or to adjacent municipalities for work. Agriculture takes place in mainly two forms: vegetable farming and decorative gardening. Both activities are economically relevant for Abidjan because they support livelihoods and produce large quantities of food and plants for urban greenery.

Vegetable farming

In 2005, a study identified three different farming areas for vegetables around the FHB airport, with the largest area counting about 800 farmers on a production area of 70 ha (Babo 2010: 102). In 2013, another study investigated 30 ha next to the airport and 35 ha on a nearby peninsula called Adjahui. In total, 299 farmers were interviewed (Wognin et al. 2013).[14] Urban vegetable farming provides livelihoods for thousands of people, especially for young men with little formal education. Besides producers, vegetable farming employs intermediate traders and market traders, many of whom are female. Long before the establishment of the cargo airport in 1974, migrants from the savannah cultivated inland valleys and plots with a high water table in Port Bouët.[15] Migrants accessed farms and garden plots through the *tutorat* with two local Ebrié villages: Ancien Koumassi and Agbabou (Koffi-Didia 2016; Shighata 2017). The northern diet was based on rice, vegetables, and leafy vegetables, rather than on yam and fish eaten by the native population of the coastal region. Different to the Ebrié, people from the north were also experienced with inland valley farming. Migrants removed the natural vegetation and reclaimed the swampy parts of the airport reserve for agriculture. At the end of the 1990s, about 80% of the urban farmers in Abidjan had a migrant background (Babo 2010), originating mainly from Burkina Faso. Over the decades, the vegetable farms and gardens became an integral part of the peri-urban fabric between the lagoon and the ocean. Work on the vegetable farms and gardens created a demand for housing in walking distance.

Initially, migrant farmers produced vegetables for subsistence and sold overproduce to the growing population of industrial workers. Later, the proximity to the cargo airport encouraged them to cultivate export crops, such as flowers, bananas, and pineapples (Nassa 2011). Current urban farms mainly provide vegetables to the markets of Abidjan, such as leafy vegetables, spices, garden eggs, okra, onions, garlic, and garden leek.

In 1979, after the introduction of the airport decree, the government compensated Ebrié villages and vegetable farmers for the loss of either customary territories or farmland. Shigata argued that the compensation entailed unintended outcomes in the local balance of power:

> [T]he natives [the Ebrié] who have recognized customary rights over space find themselves dispossessed of the land by the state. In fact, the space has been declared twice a public utility zone, which immediately places the [customary] landowners on the same footing as the migrants in land relations. Migrants can refer to legal provisions [of state law] to reject the terms of [tutorat] contracts binding them to the natives.
> (Shighata 2017: 26, quotation translated by the author)

The payment of compensations to vegetable farmers politically weakened the Ebrié. From their perspective, farmers should not have received compensation, because they were not landowners. In contrast, vegetable farmers interpreted the receipt of compensation payments as official acknowledgement of their use rights to the land. They felt encouraged to develop neo-customary leadership and new rules for land transactions within their communities. The airport decree and compensation payments thus initiated a novel neo-customary system of land tenure in Port Bouët, which weakened the Ebrié tutors but strengthened land-taking migrant communities.

When it became obvious that the state would not utilise the huge reserve for the airport in the near future, vegetable farmers emancipated themselves as landowners. They based their claim on the government's compensations, as well as on the idea that their valorisation of the land constituted ownership. Compensated farmers illegally re-appropriated their former farm plots and plantations in order to continue with cultivation. They settled on the plots, established their own councils, and sold land to house builders. As a result, especially in Adjouffou and the other three large quarters along the coast, land transactions follow neo-customary rules today.

The Ebrié villages sought to reclaim the underused airport reserve from the state by legal means, which did not succeed. They, however, also re-appropriated the expropriated land by recalling that the airport reserve was part of their customary territories and exercising customary landownership by giving out contracts for temporal rentals around the airport (Koffi-Didia 2016; Eguavoen 2021, Ittner 2023).

During the Ivorian civil war (2001–2011), the airport reserve was highly secured. The French operate a military zone next to the airport (43 BIMA) and one to the north of the airport land. Securitisation had two implications on land use in

Port Bouët. Migrant farmers without Ivorian nationality faced harassment by the military and the airport police. Farming was shifted to earlier and later hours in order to escape controls. Many migrant farmers, especially from Burkina Faso, left the country for some years and returned to Abidjan after the conflict years. Farming was technologically adapted because farmers had less time and workforce to sow, irrigate, and harvest. Despite facing difficulties and opening additional farms in other parts of the city, farmers continued to work around the airport in order to secure their plots, as well as keep up contact to their customers (Babo 2010: 109ff). Different to other areas in Abidjan during this period, they were unable to form associations and, thus, missed support from the government or development agencies. Insecure land tenure seems to have prevented farmers' social organisation (Babo 2010: 114).

The second implication of securitisation and the presence of the French military was that Port Bouët became an attractive housing zone. The pacified area experienced another influx of people, who fled from other quarters of Abidjan and rural areas. The new residents mainly became tenants. In Port Bouët, migrants felt generally safer and less harassed by xenophobic actors, because cosmopolitan neighbourhoods were an everyday experience and migrant leaders were politically powerful. Building plots and rental accommodation could easily be accessed via neo-customary land transactions and migrant networks.

In 2019, mainly young men from families of northern origin cultivated vegetable farms on the airport land, using intense gardening as a response to unemployment. They rented plots from either the Ebrié (temporal rental) or migrant farmers (commercial rental). Market traders, residents of Port Bouët, and airport workers bought their products. As these vegetable farmers were aware of the risk of eviction by AÉRIA, they prevented any erection of infrastructure on their farms and followed a short production cycle that allowed two to four harvests in a year (Kra 2019),[16] supported by the heavy application of fertilisers and pesticides. A more sustainable form of agriculture seems unsuitable under conditions of land insecurity. The study by Shighata pointed to the fact that Ebrié women, who could not directly access Ebrié customary land, rented farm plots from migrants (Shighata 2017), circumventing male-dominated, customary authority over land transactions.

During our study in 2021, we observed high levels of activity on the mixed-crop farms. On any day of the week, and at any time of the day, we saw hundreds of male and female farmers sowing, planting seedlings, and applying fertiliser, weeding, harvesting, and irrigating their fields with watering cans. A section of farms had to be abandoned for the construction of a large multi-purpose logistic centre serving the airport. Many vegetable farms on the way to Adjahui that we had documented in 2018/2019 were overgrown by unplanned housing (see next section). As a result, farmers squatted and reclaimed new land plots, including waterfronts (see Figure 6.2), while other approached Ebrié for novel rental contracts. After AÉRIA implemented the security zone around the airport wall in 2020, farmers quickly re-appropriated plots of demolished houses by removing rubble and re-establishing vegetable gardens.

106 *Irit Ittner*

Figure 6.2 Freshly reclaimed vegetable beds near the airport wall.
Source: Photograph by Ittner.

Decorative gardening

The second form of agriculture on airport land was decorative gardening. Nursery gardens selling decorative plants and trees were mainly established along the *Boulevard de l'Aéroport* and the *Route de Grand Bassam*. On the backside, these areas directly border the French military zone BIMA 43. Located on drained plots crossed by canals and small ponds, and with a rich offering of various and colourful garden species, more than 150 garden businesses decorate the entrance to Abidjan when coming from the airport. More importantly, these gardens have evolved into small ecosystems providing biodiversity and a tangible micro-climate. Migrant families from West Africa had founded these businesses and transferred them to the next generation. Land use was based on 40- to 50-year-old *tutorat* arrangements with the Ebrié of Agbabou village. Some plots were previously used as vegetable farms. Since 1979, with a stronger presence of the state around the airport, florists mainly referred to the interest of the state when discussing the insecure tenure of their gardens and the endangered future of their businesses. Large billboards indicated unequivocally that this land was part of the airport and a public utility (see Figure 6.3).

Our interview partners were afraid to lose their gardens to airport expansion. Despite the fact that they had organised a professional council, which facilitated communication with the municipal council, decorative gardeners thought they had no bargaining power because they were foreigners working on a public utility. As one old owner of a business explained to us:

> We [our garden businesses] are well located, but the land is not guaranteed. We are already here for 40 years, almost 50 years. And we live well. No problems. But it is not guaranteed. One day or the other, one can displace

Appropriation of airport reserve in Abidjan, Côte d'Ivoire 107

Figure 6.3 Billboard indicating public landownership.
Source: Photograph by Ittner.

> us. It is the day when the state arrives, as foreigners are arriving, we will be obliged to leave. . . . I paid my taxes, first to the Ebrié, then under President Ouattara, to the town hall. See, we have lost the land [the land use rights given to us by the Ebrié].
>
> (elder gardener, translated interview transcription, 23 November 2021)

This quotation shows that the businesses evolved and remained despite decades of great uncertainty. Not being able to base their use rights on customary land rights any longer, the strategy of the gardeners, therefore, was to keep the tolerated status quo, demonstrate humility, and fully comply with the rules of the airport zone. They were cautious not to contest the authority of the state, of AÉRIA, or any of its representatives (owners of gardens and gardeners, interviews, 23./24.11.2021).

To sum up this section, the farmland and the garden plots are owned by the Ivorian state and managed by AÉRIA. The long period of underutilisation for airport expansion allowed producers to re-appropriate swampy sections of the airport reserve near the lagoon. They are using the farmland until the state or AÉRIA reclaims the airport reserve for construction. There is a close historical and spatial dependency between the airport and urban agriculture in Port Bouët. Apart from the wetness of the land, which made it less attractive for housing, the legal prohibition of private landownership prevented urban development. An open peri-urban landscape persisted, with villages as the main form of housing. A massive population influx to this area began not earlier than 2011 and first manifested on the Adjahui Peninsula.

Villages accommodate up to a few thousand people. Residence is characterised by simple rural housing, as well as the lack of electricity and a potable water supply. Most villages on the public utility are tolerated by the state and formally acknowledged by the municipality. They were not equipped with adequate infrastructure by the municipality. These villages recorded a stark increase of requests for building plots and rental housing in recent years (Eguavoen 2021; Ittner 2023).

Popular urban quarters along the Atlantic coast

To the west of the airport grounds, planned urbanisation was prohibited by the airport decree as well. Along the Atlantic coast, however, spatial development took a completely different turn. *Campements* and unplanned settlements in Port Bouët date back to the 1930s. From the 1950s onwards, after the abolition of forced labour in 1947, Port Bouët faced several waves of immigration initially triggered by the construction of the Vridi Canal connecting the Atlantic with the lagoon and the establishment of industrial zones in Port Bouët. Farmers from the north of Côte d'Ivoire and former French West Africa were eventually joined by fishermen from Togo, Benin, and Ghana who searched for housing near the Atlantic, as well as by industrial workers and West African livestock traders. Four large precarious quarters that accommodate more than half of Port Bouët's population today, Adjouffou, Derrière Wharf, Jean Foly, and Gonzuageville, were initially small settlements based on early *tutorat* contracts, the illegal re-appropriation of the airport reserve after compensation in 1979, and squatting (Koffi-Didia 2016; Shighata 2017). The quarters are sub-divided into sectors and governed by neo-customary leaders (*chef de quartier*), who gained initial legitimacy through the foundation of the settlements, and local councils. Their political power is based on their central role in local land governance, as well as the dependency of the municipal government on their database and knowledge about land transfers, as well as on social-political ties in these quarters, as Chéke, Nikebie, and Gballet underline:

> The municipality is obliged to collaborate with the neo-customary actors. While refusingto acknowledge their land tenure practices in their new configuration, the governance actors rely on neo-customary systems for interventions into the management of traditional land resources
> (Chéke et al. 2018: 278f., quotation translated by the author).

Under the past municipal government, the authors observed alliance building between neo-customary actors and municipal officers' units, as well as a lack of political control by the municipality over these actors (Chéke et al. 2018).

Formal urbanisation in the centre of Port Bouët, the former location of Abidjan's wharf, was extended incrementally by private and public housing schemes. Employees of the airport were given the opportunity to buy family houses on compounds situated near the airport behind the wharf during the 1950s. Today, the Cité Météo, the Cité Policier, and the Cité Douane, as well as houses built for staff members of the Agency for the Safety of Air Navigation in Africa and Madagascar (ASECNA), constitute spatial islands on the airport reserve in Derrière Wharf with secured private house and landownership (see Figure 6.4).

Figure 6.4 Gate to the Cité Douane.
Source: Photograph by Ittner.

These spacious compounds of airport professionals are surrounded by sectors that were informally constructed on insecure building plots. The construction of public housing in the city from 1977 to 1984 was followed by an urban rehabilitation programme by the World Bank in the 1990s. *Campements*, which had evolved into larger unplanned settlements, were destroyed and evicted. Their residents moved eastwards to unused spaces of the airport reserve along both sides of the coastal road to Grand Bassam. Other evicted residents re-established smaller but also densely populated settlements along waterfronts of the lagoon behind industrial zones, the slaughterhouse, the cemetery, and the livestock market.

While unplanned housing in Port Bouët first served as accommodation for West African migrants, these residential areas later became popular among *Abidjanais* of all origins due to their proximity to industrial jobs and low rents. Because the acquisition of legal landownership was impossible on the public utility, Port Bouët evolved into a municipality dominated by low-cost, precarious housing with a cosmopolitan population. When the municipal system was introduced in 1978 and population pressure increased, settlements in Port Bouët experienced an influx of people from the lower and middle classes, of foreign and Ivorian origin alike, who bought building plots under the neo-customary regime, rented housing, or bought housing in the few formal housing schemes outside the airport reserve, such as in the Cité Atlantique. New buildings in Port Bouët centre also included military camps and student dormitories. From 1990 to 2016, only two more housing schemes were built, including '48', constructed by ASECNA for airport workers (Koffi-Didia 2016).

While Derrière Wharf's growth was spatially limited by the *Boulevard de l'Aéroport* and the airport grounds, Adjouffou sprawled to the north and to the west right up to the security wall of the airport. The subsequent national governments and District of Abidjan tolerated informal housing practices, although the residents of Derrière Wharf and Adjouffou were constantly reminded that they lived illegally on the airport reserve and would need to leave one day. An urbanist told me about observations during his internship with the technical service of Port Bouët municipality in 2011:

> *I worked on the restructuring of the quarter Derrière Wharf, just behind the military camp. The first thing I noticed was that residents had already been compensated. They lived there for long. They* [the District of Abidjan] *had to evict them because of airport expansion. In an official letter, which was published in a newspaper, it was said that these people would be evicted because they lived on the airport reserve. They received this letter in 1960 or 1965. Until 2011, they were not evicted. No airport expansion project was implemented. You find yourself in a situation where residents know that they are illegal. But nobody comes and tells them to leave. There were even resettlement zones which they never went to.*
> (Mr Konan, translated interview transcription, 19 March 2019)

As this quotation illustrates, illegal housing on the airport reserve was tolerated for decades by state authorities, who showed a *laissez-faire* attitude to the situation on the ground. Housing on the underused airport reserve was supported by the municipal council of Port Bouët and its mayor, Madame Hortense Aka Anghui, who was in office from 1980 to 2017. The council applied a regulation from 1980 that allowed local governments to permit housing on public land within their municipality by the allocation of temporal housing permits. Housing rights were registered by the councils in the quarters.[17] Unplanned settlements became formally integrated urban quarters, and their residents enjoyed a temporal legality. The municipal council could govern with the support of the neo-customary authorities and provide municipal services for the residents.

During the early 1990s, when the welcoming immigration policy of Felix Houphouët-Boigny ended, the government of Bedié, as well as the following governments under Robert Gueï (1999–2000) and Laurant Gbagbo (2000–2010), openly promoted an ultra-nationalist ideology (Ivorian nativism) and xenophobia. Laws barred foreigners from legal landownership and introduced a very restrictive nationalisation law (Rosa-Luxemburg-Stiftung 2020). Encouraged by these public discourses, groups of Ivorian *Abidjanais* protested against the neo-customary councils. They argued that foreigners, who had illegitimately appropriated Ivorian public land, should not be selling it to Ivorians. Protesters squatted other parts of Port Bouët and began to build housing. They were joined by young Ebrié who had also experienced a land and housing shortage. Their justification differed in the sense that they underlined their customary rights to the land in Abidjan as autochthons (Koffi-Didia 2016).

In 2012, the total population of Adjouffou alone was estimated at 100,000 inhabitants, many of them living under slum-like conditions in low-quality housing

(ONU-Habitat 2012). The high population of Adjouffou and other precarious quarters in Port Bouët became a political variable in itself (ONU-Habitat 2012), because house owners and tenants made up the highest numbers of voters within the municipality.

For humanitarian reasons, the high number of residents alone actually forbids an urban renewal policy based on demolition and displacement. The municipality lacks land resources and money to create additional housing. The challenge of homelessness is especially felt after evictions, when residents expect the municipal council to solve their housing problem, at times by directly attacking the town hall.[18]

Legally secured but forceful evictions of housing patches and smaller unplanned settlements were conducted by the District of Abidjan, who justified these piecemeal demolitions and displacements with either the construction of the international coastal road to Accra, disaster risk reduction, or the city cleaning programme.[19] During the time of study, demolitions along the *Route de Grand Bassam* were announced for the construction of the metro that will link the airport with different parts of the city. Evictions finally took place in July 2022.

The municipal council faced emergencies when thousands of inhabitants became homeless due to evictions along the Airport Road (Moussakro in 2011), along the Atlantic Beach (parts of Derrière Wharf, Adjouffou, Jean Foly, Gonzuageville, Anani in 2013/14), and along the waterfront of the Bay of Biétry (Quartier des Éleveurs in 2017, Abbatoir in 2018).[20] These evictions created intra-urban flows of poor tenants searching for housing. This resulted in the densification of people in existing precarious housing, as well as in higher entry cost to rental contracts.

It also contributed to the rapid evolution of Adjahui, a new informal settlement on the airport reserve, which already accommodated more than 60,000 residents in mid-2018. The respective peninsula was mainly covered by three villages, palm plantations, farms, and bush land before urbanisation was set in motion by the eviction of Moussakro in 2011, by the emergency financial support for displaced people paid by the municipality, as well as by the government's announcement of an airport city, which would change the spatial fabric on the airport reserve (Eguavoen et al. 2020; Eguavoen 2021).[21]

The analysis shows that population pressure and eviction policy in the context of legal pluralism in tenure and the underused airport reserve resulted in a concentration of people with low incomes in precarious quarters of Port Bouët. Urban policy at the metropolitan scale pushed poor households into the urban periphery and public spatial niches, where no secure land title could be acquired, while the municipality aimed at securing housing and temporal residential rights for them. Today, the massive population number on the reserve hinders the airport city project. Implementation could only be enforced by evictions and would face massive resistance by the house owners, the councils of the quarters, and the tenants. It is also unlikely that the municipal council of Port Bouët will push the project as much as the national government and the District of Abidjan do.

At the same time, precarious housing also mushroomed in the municipality. Inhabitants of evicted settlements re-established these along the road to Grand Bassam. Their offspring grew up in the city. They are *Abidjanais* of the second or third generation. Their home country is Côte d'Ivoire, even though Ivorian nationality was denied to them by law. Some inherited land plots and houses from their families.

The eviction of Adjouffou Premier Arrêt – resistance and return

Keeping these two spatial evolutions in mind, we return to the security zone in Adjouffou, where about 2,000 households and 156 landlords were initially given 48 hours to leave the sector in 2020. Shortly after the call by the Ministry of Transport to enlarge the security zone of the airport, public officers arrived and marked the buildings with painted crosses and the letters *AD* (French: *à détruire*, to demolish) in order to announce their eviction. Two members of the committee of the residents of Adjouffou Premier Arrêt, an informal group representing tenants and landlords affected by the eviction, recalled the events during an interview in December 2021. We documented roads of deserted buildings with disassembled features and partly demolished walls before arriving in streets where people had re-appropriated the evicted houses. For Adjouffou, a popular quarter with high population density and many businesses, it was remarkably calm. The men said that everybody in Premier Arrêt knew that land use, as in Adjouffou in general, was temporary and would end when the state reclaimed the airport reserve. Our interview partners and other residents around the airport communicated their sincere doubt about the fatal journey of the boy, which served as justification for the extension of the security zone, and the eviction. Some suspected that the story was made up by either the district or state authorities. Others asked how the boy could know the place well enough to sneak into the airport grounds when he originated from a far municipality. According to them, professional security measures behind the airport wall prohibited any access. From their perspective, the extension of the security zone was part of a piecemeal strategy by the state to reclaim public land of the airport reserve, as Adjouffou and the other urban quarters could not be evicted without causing a humanitarian disaster. Our interview partners explained why they had protested against their eviction:

> *If there is to be an eviction, I think it should be done in a structured way, with support and a period to allow residents to find another place to go. Because this is a property of AÉRIA. It is state land. The state, however, has no right to force a population to move within 48 hours and without accompanying measures. This is an act of force.*
> (committee members of the residents of Adjouffou Premier Arrêt, translated interview transcription, 2 December 2021)

Thus, the tenants did not claim ownership of their residential place or contest the public domain. They were ready to leave but insisted on the lawful implementation of the displacement and government support to at least partly compensate for the loss. Of course, many other residents in Adjouffou hoped to avert evictions by public protest. A wide coalition of local organisations, associations, and leaders called for a demonstration. A sociologist at the University of Cocody described some dramatic scenes he observed during the emotional demonstration in Adjouffou, when some people stood on roofs shouting that they would rather die than leave.[22] Schoolchildren begged through tears that they wished to continue their school year. TV reporters broadcast the scenes. Because of the protests, the date of demolition was postponed by another two weeks. Tenants used this period for house hunting, packing, entering new rental contracts, or moving to the homes of their relatives. Landlords made sure they left nothing of value behind. They dismantled the buildings. Zinc roofs, windows, and doors were taken away. Providers cut electricity and water supplies. Some neighbours, including our interview partners, addressed letters to various Ivorian ministries, as well as to international politicians:

We wrote to embassies, to the different chancelleries [Germany, USA, France, USSR], to human rights NGOs, to the Ministry of Solidarity, to the Ministry of Justice, to the Ministry of Education. Because there were schools that were to be demolished in the middle of the school year, we had to get the pupils to go out and cry in front of the press, saying that they wanted to go to school. . . . We even wrote to the representatives of the African Union and the EU [European Union].
(committee members of the residents of Adjouffou Premier Arrêt, translated interview transcription, 2 December 2021)

While most of the letters remained unanswered, Adjouffou residents succeeded to create political pressure and to draw the attention of the media even at an international level after their family members in Great Britain protested in London against the eviction, where the Ivorian president Alassane Quattara was attending an international meeting. He was confronted with uncomfortable questions by the international press about what was going on in Côte d'Ivoire. From the perspective of the residents, critical international media reports prevented the demolitions.

Residential buildings stood empty for more than a year, which encouraged residents who had not found affordable accommodation in the meantime to contact their landlords and return to their former housing units. Housing conditions, however, had deteriorated, and arguments between landlords and tenants arose about the payment of the cost for repairing buildings and setting up new water and electricity connections.

To sum up this section, Premier Arrêt is the section of Adjouffou that directly borders the airport wall and was affected by the implementation of the security zone in 2020. Despite announcing evictions and making residents leave, public protest and media attention created a political climate that made it very difficult for the state to implement demolitions. Unused and undemolished buildings were partly re-appropriated by former residents and house owners. Even though the sector is very small in relation to all land that was illegally occupied on the airport reserve, the state and District of Abidjan had not enough authority and power to reclaim the land plots on the public reserve.

Discussion and conclusion

The high economic ambitions of the Ivorian government are informed by the role of Abidjan in its golden age during the 1970s and 1980s, when the metropolis served as an economic hub of West Africa. Though national economic growth rates are considerable, the number of passengers travelling through FHB airport in comparison to other African international airports is still low. Murtala Muhammed International Airport in Lagos, with more than 7 million air travellers (before Covid-19), stands as the biggest rival in the fierce competition for the West African aviation hub.

The Ivorian government considers air transport essential for the economic growth of the country. It introduced an airport modernisation programme in 2014. The programme is motivated by the need to expand the aviation sector in order to not be left behind in the global competition as a business hub, as well as in order to support other economic sectors of the nation. Abidjan already owns a large and expanding international harbour, though operations have faced problems because of piracy in the Gulf of Guinea. The FHB airport received a new cargo airport. New logistical enterprises are set up on the airport land. Airport expansion and the airport city face difficulties in funding and implementation on the ground because the airport reserve, a public utility, has been occupied by various other users for many decades. Transformation in land use would result in massive changes of the social-economic fabric in Port Bouët and political instability. The implementation of any eviction and final demolition for the establishment of new infrastructure is a political challenge for the government.

Practices of urban citizenship and private investment turned the underused airport reserve into an intensely used urban landscape with important agricultural uses and settlements (villages, unplanned settlements, and urban quarters) that host about half of the population of Port Bouët. The spatial co-evolution with the airport included two different pathways leading to a peri-urban landscape in the east and north on the one hand and a dense urban fabric, especially in the west and south, on the other. These pathways were enabled by different landscape features, legal pluralism (land law, customary and neo-customary tenure) and by the competing development vision and contested leadership of the state and the municipality over the airport reserve (state-led development of airport-related uses, business and tourism site versus stepwise integration of unplanned settlements into the municipal structure). It was also enabled by the historical evolution of Port Bouët into a periphery of Abidjan with unplanned, low-cost housing. The growth of local informal livelihoods and precarious forms of housing, as well as the exclusion of migrants from legal landownership, entailed the evolution of a dense network of social associations and new forms of local land tenure governance and political leadership, as well as of vivid, cosmopolitan, and defensive communities with high voters' base, which gained political weight due to their high population numbers. The Ouattara government is intending to reform the nationalisation law, which might allow more people to legally access secure land titles after the receipt of Ivorian citizenship.

Perhaps it was naive to think that about 100 km² of land near the growing metropolis would remain barren land, or that temporary uses, practices, and claims for landownership would not perpetuate. Acts of urban citizenship hinder formal urban planning and the mid-term implementation of the airport city. House owners and tenants, who see their interests endangered by the airport city, learnt by experience which resistance strategies had worked well over the past decades. They mobilise into community-based organisations and NGOs and search for media attention. They play the customary/neo-customary or 'poor people's area' card and seek patron–client relations with local politicians. They protest on the street, file lawsuits against authorities, and communicate that they are quite aware of their right to accommodation and being in the city. Facing a neglecting *laissez-faire* policy, civil disobedience, and the ignorance of eviction notices proved to be the most successful resistance strategy by residents. After the Ivorian civil war, the national and district governments consequently began to implement the rule of law in order to reclaim their authority and to sanction illegal squatting on the public airport land. The present political means to do this, especially demolitions and forceful evictions, should be questioned because they create humanitarian emergencies and exacerbate the housing crisis. Now, unplanned settlements provide the only response to the housing crisis in the low-price segment of Abidjan (Eguavoen 2021). If evicted places are not fully occupied by public authorities, people tend to reclaim them. It seems that the state and the District of Abidjan have lost some rounds in this regard in Premier Arrêt and elsewhere.

The employment generated by the future airport city was estimated at between 20,000 and 35,000 jobs. Most jobs will be formal, better employment and probably be better paid than existing jobs in the local economy. Evictions, however, also destroy jobs and economic assets. Farmers and gardeners acknowledged the role of AÉRIA as quasi-private landowner and knew that they would need to leave the land as soon as the state claimed it back. Despite putting the land into value for many decades, they had succeeded neither in securing their land use rights nor in building strong political alliances that would help them continue with their livelihood in the event that their gardens and farms were needed for airport expansion.

Therefore, which policies would be promising on the way to urban development, the well-being of the population, and social-economic sustainability? Of course, airport expansion should be critically reflected upon as well with regard to the number of expected air travellers and needed capacities. It seems that the airport city serves rather as a justification to reclaim land for formal urbanisation that promises high economic gains for the state, the district, and developers. A policy option would be to tolerate the existing farms and settlements until 2030, when the concession with Egis runs out. Inhabitants around the airport would need to leave only land plots that are urgently needed for airport expansion.

After 2030, the size of the airport reserve could be drastically reduced by a new government decree. The public utility status of the cut-off land could be transferred to the municipality. This would be important in order to prevent land speculation and gentrification. Urban development outside the reserve would then continue under the responsibility of the Port Bouët municipal council in cooperation with

116 *Irit Ittner*

the district government. This would enable the improvement of precarious housing under public housing schemes for households with low incomes and other housing development. Competition between the district and the municipality, however, over the control of land resources, and between the municipality and the neo-customary leaders in Port Bouët over political power, would probably undermine such a policy option.

Acknowledgements

The Fritz Thyssen Foundation funded the project 'Everyday entanglement of social-economic extremes and negotiations in anticipation of development-induced displacement in Mumbai and Abidjan' (2020–2023, 10.20.2.003EL). The foundation also funded open access. I would like to thank Sneha Sharma and Dishani Senaratne for the review of earlier versions, and Oscar Eguavoen for designing the map.

Notes

1. For the concept of airport land, see Sharma et al. (2024).
2. The French term *foreigners* (*les étrangers*) includes different notions, such as not having Ivorian nationality or, more generally, not originating from this place. This chapter, however, refers strictly to the legal notion of the term 'foreigners'.
3. 04.04.2019 AIP. They lived in an unplanned settlement called Plaque Air Ivoire; 25.08.2011 Nord-Sud; 10.05.2015 AIP.
4. Census data from 2022.
5. The Ivorian civil war and post-electoral crisis (2001–2011) marked a political turning point. For a comprehensive country overview, see Miran-Guyon (2017).
6. Kouassi (2021) divides them as follows: 1956–1960 colonial period, 1960–1969 period of post-independence, 1969–1981 period of infrastructural development and expansion, 1981–1999 period of economically difficult years, 1999–2011 period of decline (due to the civil war and post-electoral crisis), since 2011 period of revival.
7. 16.09.2018, newsaearo.info/connectionivoirienne.net.
8. Other policy measures included the introduction of an easy online procedure for visa application, American TSA (Transport Security Administration) certification, as well as the construction of a new runway, which permitted the first landing of an Airbus A380 (the largest civil aircraft of the world) in Sub-Saharan Africa. FHB is also the leading African airport with regard to the Airport Carbon Accreditation Level, which is 3+/neutrality. Following the open-sky policy, the Ivorian government signed contracts with 25 international airlines that led to a rapid increase in the number of destinations. In 2018, a milestone was the introduction of a direct connection between Abidjan and New York. It is offered by Air Côte Ivoire three times a week in alliance with Ethiopian Airlines. In recognition of these revitalisation measures, FHB airport received the Africa Routes Marketing Award 2018 for airports with passenger numbers under 4 million travellers. Its efforts were also recognised in 2019 by the Airports Council International, who gave the Airport Service Quality Award to Abidjan in the 'Most Improved Airport' category (25.02.2014 news.abidjan.net; 01.03.2014 news.abidjan.net; 13.02.2015 news.abidjan.net; 27.08.2015 Le Diplomatique D'Abidjan; 31.12.2015 news.abidjan.net; 31.04.2021 Le Patriote; newsaero.info; abidjan-airport.com).
9. According to Shighata (2017: 58, 79), the airport land extends from latitude 5°00–5°30 and longitude 3°50–4°10. It is limited by the municipalities of Marcory and Koumassi to

the north and the Atlantic to the south, as well as by Grand Bassam (a city and prefecture of the District of Abidjan) to the east.
10 Following Egis in 2019, the shares of the airport are currently distributed among Egis (34.5%), SEGAP (African Investment Managers and Afripar, total 27.1%), the Ivorian government (10%), and private investors (29.1%).
11 For example, the professional training centre for airport workers (*Centre des métiers de l'aviation*), the sports club *Aéro-Club D'Abidjan*, and airport transfer and parking, as well as logistic companies.
12 This status is usually determined by age, gender, class, and/or belonging to an age class. Chauveau and Colin (2010), Shighata (2017: 76f.).
13 In typical Ebrié parlance, Ebrié *chefferies* refer to 'installed migrant communities on the land'.
14 The agricultural site in Adjahui was completely transformed into a dense settlement by 2018 (Eguavoen et al. 2020; Eguavoen 2021).
15 The passenger airport was established in 1939 at the same place.
16 These findings are supported by Wognin et al. (2013), who underline that predominantly uneducated young men made up most of the vegetable farmers.
17 Law 85–582, 29.07.1980. The law permitted the municipal council to parcel the land and request fees and taxes.
18 02.07.2018 infodrome.
19 In 2012, l'*Organisation des Secours* (ORSEC), the government organisation for disaster risk reduction, listed eight quarters of Port Bouët as 'risky zones' (Adjouffou, Gonzuageville, Jean Foly, Anani, Abrogoua, Tavito, Petit Bassam, and Vridi-Canal), where no housing was permitted in order to protect people from potential hazards, such as flooding. Residents were given orders to leave before these quarters, or parts of them, were evicted and demolished in 2013/2014.
20 The eviction was justified with potential health hazards due to open sewage from the slaughterhouse, as well as the location of the livestock holders' homes under high-voltage lines.
21 The eviction created a housing emergency. Displaced households used financial support for the purchase of land plots on the peninsula (additional to their own resources). The announcement kicked off land speculation.
22 R. M. Chéke, personal conversation, December 2021.

References and sources

Babo, A. (2010). 'Sociopolitical crisis and the reconstruction of sustainable periurban agriculture in Abidjan, Côte d'Ivoire'. *African Studies Review* 53(3), 101–120.
Baégas, R. and R. Marshall-Fratani (2007). 'Côte d'Ivoire. Negotiating identity and citizenship'. In *African guerrillas. Raging against the machine*. M. Boas and K. C. Dunn (eds.). London: Lynne Rienner, 81–111.
Bredeloup, S. (2003). 'La Côte d'Ivoire ou l'étrange destin de l'étranger'. *Revue Européenne des Migrations Internationales* 19(3), 85–113.
Chauveau, J.-P. and J.-Ph. Colin (2010). 'Customary transfers and land sales in Côte d'Ivoire. Revisiting the embedded issue'. *Africa* 80(1), 81–103.
Chéke, M. R., K. C. S. Nikebie, and R. L. Gballet (2018). 'Acteurs sociaux, gestion foncière périurbaine et dynamiques sociales à Abidjan (Côte d'Ivoire)'. *KASA BYA KASA – Revue ivoirienne d'anthropologie et sociologie* 37, 269–285.
Eguavoen, I. (2021). '"We do the social." Deal-making by non-accredited estate agencies, small-scale investors and tenants around low-cost rental housing in Abidjan, Côte d'Ivoire'. *Afrika Focus* 34(2), 183–212.

Eguavoen, I., P. Attemene, F. Kouame, E. K. Konan, C. A. Madhy, and K. Gleisberg-Gerber (2020). 'Dernier refuge ou presqu'île d'opportunités? Démographie et conditions de vie à Adjahui-Coubé, un habitat spontané à Abidjan'. *ZEF Working Paper* 187. Bonn: ZEF.

Fall, M. and S. Coulibaly (eds.) (2016). *Diversified urbanization. The case of Côte d'Ivoire.* Washington, DC: World Bank Group.

Isin, E. F. (2009). 'Citizenship in flux. The figure of the activist citizen'. *Subjectivity* 29, 367–388.

Ittner, I. (2023). 'Emerging disputes over land and leadership in urban villages on the airport reserve in Abidjan'. *Afrique Contemporaine* 275(2), 175–201.

Koffi-Didia, M. A. (2016). 'Port-Bouët, une commune au cœur des processus de recomposition spatiale à Abidjan (District d'Abidjan)'. *Syllabus* 7(1), 93–114.

Kouassi, K. K. S. (2021). *Avenir du transport aérien international en Cote d'Ivoire à l'horizon 2065.* Thèse de géographie pour l'obtention du grade de Docteur. Université Félix Houphouët-Boigny, Abidjan.

Kra, K. V. (2019). 'Maraîchage intra-urbaine à Abidjan et Bouké (Côte d'Ivoire). Entre économie spéculative et dimension socio-culturelle des acteurs'. *Nzassa* 2, 343–354.

Mairie d'Abidjan (1963). *Abidjan. Capitale de la République de Côte d'Ivoire.* Abidjan: Société des Éditions Paul Bory.

Marshall-Fratani, R. (2006). 'The war of "who is who": Autochthony, nationalism, and citizenship in the Ivoirian crisis'. *African Studies Review* 49(2), 9–43.

Miran-Guyon, M. (2017). 'Côte d'Ivoire. The return of the elephant? Introduction'. *Afrique Contemporaine* 263/264, 9–22.

Nassa, D. D. A. (2011). 'Le commerce à une frontière paradoxale'. *L'exemple de l'aéroport d'Abidjan*. Retrieved 24.05.2024. https://shs.hal.science/halshs-00580335

ONU-HABITAT (ed.) (2012). *Côte d'Ivoire. Profile urban de Port- Bouët.* Nairobi: ONU-HABITAT.

Raynaud-Matheis, F. (1962). *Elfenbeinküste*. Bonn: Kurt Schröder.

Rosa-Luxemburg-Stiftung (2020). 'Elfenbeinküste. Für Kaffee und Kakao'. In *Atlas der Staatenlosen. Daten und Fakten über Ausgrenzung und Vertreibung.* Rosa-Luxemburg Stiftung (ed.). Paderborn: Bonifatius, 40–41.

Sanogo, A. and L. Doumbia (2021). 'Land tenure and public infrastructure. Airport building in Côte d'Ivoire and Senegal'. In *Land tenure challenges in Africa. Confronting the land governance deficit.* H. Chitonge and R. Harvey (eds.). Cham: Springer, 203–225.

Sharma, S., I. Ittner, S. Mingorría, I. Khambule, and H. Geschewski (2024). 'Contested airport lands in the Global South'. In *Contested airport land. Social-spatial transformation and environmental injustice in Asia and Africa.* I. Ittner, S. Sharma, I. Khambule, and H. Geschewski (eds.). Abingdon: Routledge.

Shighata, C. B. (2017). *Les systèmes fonciers coutumiers face à l'urbanisation: Recompositions sociales et conflits fonciers, Abidjan (Côte d'Ivoire).* Bielefeld: ÉUE.

Wognin, A. S., S. K. Ouffoue, E. F. Assemand, K. Tano, and R. Koffi-Nevry (2013). 'Perception des risques sanitaires dans le maraîchage à Abidjan, Côte d'Ivoire'. *International Journal of Biological and Chemical Sciences* 7(5), 1829–1837.

Zina, O. (2017). 'National reconciliation and constitutional reforms in Côte d'Ivoire'. *Afrique Contemporaine* 263/264, 23–36.

Media sources

25.08.2011. 'San Pedro: des habitations sur la plate-forme, la mairie accusée'. *Nord-Sud*. Retrieved 10.05.2021. https://news.abidjan.net/h/408651.html

25.02.2014. 'Construction de l'Aérocité. Plusieurs milliers de sans abris annoncés à Port Bouët'. *Le Démocrat*.

01.03.2014. 'Cérémonie de lancement des traveaux et d'extension d'aérogare de fret de l'aéroport international Félix Houphouët-Boigny. Allocution du premier ministre'. Retrieved 06.07.2016. http://news.abidjan.net/h/490580.html

13.02.2015. 'Ministère de la communication Affoussiatou Bamba face à la presse, ce matin'. Retrieved 06.07.2016. http://news.abidjan.net/h/525994.html
10.05.2015. 'La mairie va déguerpir les occupants illégaux dès juin en vue de l'agrandissement de l'aéroport de San Pedro'. *AIP*. Retrieved 10.05.2021. https://news.abidjan.net/h/551298.html?fb_comment_id=1104790826203353_1104834986198937
27.08.2015. 'Aéroport d´Abidjan. L´aérocité, un grand reve qui risque de ne pa se réaliser'. *Le Diplomatique D´Abidjan*.
01.10.2015. 'Aérocité d´Abidjan. Vers une logique de rentabilité'. Retrieved 06.07.2016. http://news.abidjan.net/h/564945.html
31.12.2015. 'Aéroport d´Abidjan. De gros investissement prévus pour une meilleure compétivité'. Retrieved 06.07.2016. http://news.abidjan.net/h/578205.html
02.07.2018. 'Port-Bouët/Déguerpissement de l´abattoir. Des individues en colère saccagent les locaux de la mairie'. www.linfodrome.com/societe/39924-port-bouet-deguerpissement-abattoir-des-individus-en-colere-saccagent-les-locaux-de-la-mairie (retrieved 10.06.2021).
04.04.2019. 'Le quartier "Plaque Air Ivoire" de San Pedro déguerpi'. *AIP*. Retrieved 10.02.2020. https://news.abidjan.net/h/655215.html
30.04.2021. 'Coopération franco-ivoirienne. Amadou Koné et Bruno Le Maire signent pour l´extension de aéroport d´Abidjan'.
31.04.2021. 'Gaoussou Touré présente ´Aérocité´ aux investisseurs. Un projet de plus de 1.000 milliards FCFA'. *Le Patriote*.
'Extension de l'aéroport d'Abidjan en Côte-d'Ivoire pour 8 millions de passagers d'ici 3 ans'. newsaero.info/connectionivoirienne.net
'Routes Africa Marketing Award 2018'. www.abidjan-airport.com/routes-africa-marketing-award-2018/ (retrieved 02.06.2021).

7 The Durban Aerotropolis

Emerging and underlying territorial contestations in South Africa

Isaac Khambule

Introduction

Airports are increasingly becoming new centres of economic concentration because of the rise in air travel, aviation-related businesses, tourism, and commercialisation of airport spaces. This trend is attributed to the globalisation process that has spurred global connectivity through air travel as the most expedient mode of transportation and globalisation of cities (Goetz & Graham 2004). This process has facilitated increased trade, cooperation, investment, and job creation through airport spaces. In the African continent, the aviation industry provides more than 6.9 million jobs and contributes $80 million towards gross domestic product (GDP), with South Africa and Ethiopia emerging as the leading aviation centres (Netherlands Enterprise Agency 2015). This has fuelled the already-growing number of aerotropolis developments by cities across the globe to maximise the competitive advantage offered by airports. This is achieved by establishing export-oriented special economic zones (SEZs) in airport hubs as a new phenomenon in urban and spatial development.

Kasarda (2006) argued that there is a rapid transition of airports from standard aeronautical infrastructures to business-oriented and multifunctional projects. This ongoing transition follows the rising global demand for aviation-related and aviation-dependent businesses that build economic connectivity, thereby spurring growth and employment opportunities for host territorial units. However, this process includes territorial contestations because of various public and private interests in the development landscape. Within aerotropolis development, these contestations emerge because airports are underpinned by relational socio-spatial and political processes that impact and shape global flows, urbanisation, and territorial governance interface (McDonough 2014). This is because airports are built on land, which is one of the most contested resources globally.

In the Global South, noticeable contestations are emerging because the building and expansion of airports occurs in contested land underpinned by the displacement and dislocation of people and communities from communal and traditional land (Huka 2021). Such contestations can also be found in Bond's (2015) criticism of the Durban port-petrochemical complex. Bond (2015) classifies this development as a neoliberal development arena of social, economic, and environmental violence

DOI: 10.4324/9781003494966-7

for dismantling small-scale farmers and neighbourhoods and contributing to climate change. Thus, despite the debates on the sharing of socioeconomic benefits (employment and increased economic activities) and costs of such megaprojects, these underlying contestations affect the conservation of traditional societies and threaten democratic development. These contestations centre on the importance of land use, the diverse territorial interests of different stakeholders, and the impact of such megaprojects on health and environment in the developing societies.

Most of the literature on contested territorial airport spaces focuses on the contestation between airport developers and the dislocation of marginalised groups from traditional and communal land. This study extends this focus by concentrating on the marginalisation of territorial spaces and administrative units in the form of a South African metropolitan municipality and rural and small-town municipality in the governance of megaprojects and the sharing of economic benefits and costs.[1] Methodologically, this chapter uses secondary qualitative data on the Durban Aerotropolis project and place-based development in South Africa. Extending the analysis to a broader perspective on territorial institutions is important in understanding the institutional factors between socioeconomically diverse local municipalities and their respective governments. Essentially, this chapter is concerned with the following question: What are the emerging and underlying territorial contestations within the Durban Aerotropolis between two municipalities that surround the project, namely, eThekwini Metropolitan Municipality and iLembe District Municipality in KwaZulu-Natal Province, South Africa? The Durban Aerotropolis is significant because it is strategically located between two municipalities that hope to share the economic spoils of this megaproject in a political context, in which each locality would use its competitive advantage. Essentially, the decisions made by the governments of both municipalities will influence economic convergence or divergence within the aerotropolis region in KwaZulu-Natal.

This chapter is structured as follows: The following section reviews the literature on the rise of aerotropolis as an economic development strategy and its impact and challenges. The third section focuses on some territorial contestations of airport land development and points to the lack of literature on territorial contestations of benefit sharing between two territorial units in the form of municipalities. The fourth section introduces the Durban Aerotropolis in South Africa and the progress of this important megaproject. The fifth section delineates the emerging and underlying contestations between the two municipalities. The sixth section discusses territorial contestations in the Durban Aerotropolis in relation to other global developments and presents some policy implications for aerotropolis development. The last section concludes the chapter with policy implications for new governance frameworks.

The rise of aerotropolis as an economic development strategy

Traditionally, airports are aviation spaces for moving people, goods, and services from one geographical area to another (Button 2010). However, with fast-paced globalisation, airports have increasingly been playing a leading role in the global connectivity of cities within countries and outside geographical, national, and continental boundaries. Contrary to the fact that some authors (Goetz & Graham 2004; Kasarda 2011) have identified airports as leading contributors to globalisation, it can be argued that the globalisation process significantly contributed to the rise of airports in the commercial space. This can be attributed to the demand for speedy connectivity between different geographical locations for business, tourism, and leisure purposes. This demand has fuelled the transition of airports from basic structures to those surrounded by commercial and industrial developments to maximise airport services and increase the global competitiveness of city regions and economies. This development strategy caters to the aeronautical and commercial functions of the modern global economy (Kasarda 2006). Airports are increasingly transitioning from standard aeronautical infrastructures to business-oriented, multifunctional projects. This transition follows the rising global demand for aviation-related and dependent businesses that build economic connectivity, thereby spurring growth and employment opportunities for host territorial units.

In response to the increased connectivity and economic opportunities offered by airports, a gradual 'aerotropolis development' has occurred. Aerotropolis has emerged from the increased use of airports and airport spaces as modern drivers of business development, economic activities, and urban development through a series of aviation networks and connective infrastructures that provide brokerage for business networks and economic agglomeration effects (Kasarda 2013). Charles et al. (2007) argued that aerotropolis development promotes service industries to fulfil the needs of airport users. This follows the zoning of industrial airport land to favour export-oriented businesses, with airport land being the epicentre of economic development. Utilising the growth and volume of airports, companies have built hotels and other retail businesses within airports to maximise traffic flow and tap into the needs of travellers (Schaafsma et al. 2008). These developmental progressions have added to the commercialisation and integration of airports into the economic system of attracting travellers and caused airports to compete for more travellers. Essentially, the benefits attached to the connective infrastructure play a leading role in determining airport usage.

It is vital to understand the emergence of aerotropolis development in relation to the vision of the host territory, that is, the municipalities that host the aerotropolis. The establishment of an aerotropolis is pioneered by a city that uses its proximity to the airport as a competitive advantage and a means of brokering other economic activities and economies (Banai 2017). In the bid to benefit from the existing airport, cities use their economic power to lobby for the extension of airports and to build and cluster industries in proximity to the airport. Based on this observation, Freestone and Baker (2011) defined an *aerotropolis* as a city purposefully developed around an airport to serve as the main centre of economic activity. Such an airport performs multipurpose functions that promote and attract investment around the airport to build aviation-dependent businesses. Importantly, attracting investment through agglomeration effects and the lower cost of doing business in airports also positively impact local employment opportunities. Expanding export-oriented industries will increase economic activities and generate employment opportunities for locals, thereby attracting additional investment to create a functional aerotropolis.

In taking advantage of their proximity to airports, cities pitch their territories as gateway ports for doing business. It is argued that aerotropolis development is supposed to introduce a new development phenomenon through cities driven by 'manufacturing plants reliant on airborne inputs . . . and major road and rain infrastructure connected' to the airport (Charles et al. 2007: 1009). This has gradually led to the housing of SEZs within airport jurisdictions to facilitate export-oriented industrial and logistic parks to increase the agglomeration effects of aerotropolis development. This signals a shift away from the diverse freight-driven agglomeration of SEZs predominately focused on seaports as the main hubs to the emerging use of airport-centric SEZs to benefit from incentives such as tax breaks that attract foreign investors. The increasing incentives offered through SEZs are beneficial for existing and new enterprises and provide multiple opportunities for building new export-oriented industries in proximity to airports. Bridger (2017) observed that businesses adjacent to airports in the United Kingdom, such as Manchester, Blackpool, Luton, and Newquay, received substantial subsidies linked to SEZs.

An airport such as Indianapolis International Airport has strategically positioned its port as a regional economic node through recent investments and upgrades (Kasarda 2011). This has given birth to entrepôts (port city) and relational cities. Sigler (2013: 612) defines *relational cities* as 'those constituted through globally critical flows of capital, goods, and ideas, and whose economies are dedicated to intermediary services such as offshore banking, container- and bulk-shipping, and regional re-exportation'. Modern cities driven by global capital flows, such as Singapore, Hong Kong, Doha, Dubai, and Panama City, act as financial brokers in one way or another. Gateway cities and entrepôts are almost similar to relational cities as they 'are found eccentrically at one end of a fan-shaped network, connecting the global economy with a regional economic matrix' (Sigler 2013: 612). Cities such as Amsterdam play an essential role as gateway cities and a leading entrepôt, as evident in the role played by the city in the historical trading system of the Dutch Empire. Today, airports are the backbone of the emerging role of these cities in the global economy, necessitating the expansion of airports to serve the growing base. This role denotes the importance added by entrepôts in the competitive advantage of brokerage cities and improving regional development prospects.

Territorial contestations of megaprojects

Governments are perpetually criticised for land use and planning driven by the neoliberal order that prioritises the interests of capital in the development process. Literature on 'global land grab' significantly shows how neoliberal policies and principles have positioned corporations to grab land and resources through violent dispossession and displacement of communities (Borras & Franco 2012). Da Silva et al. (2020: 1301) criticise the planning in Goa for emerging 'as a terrain of struggle between a state-capital nexus seeking to dispossess and convert land, and an organised citizenry seeking to use planning for alternative purposes'. In

Brazil, the Bolsonaro government, working with corporations, has been invading Amazon land occupied by traditional communities to access resources, despite its negative ecological impact on nature. These land grabs are undertaken by using the state as an instrument of the neoliberal agenda in the wake of failures to dismantle the state's and traditional communities' control of valuable land. Land grabs between the state and corporate entities are also evident in Xolobeni, South Africa, despite the court's ruling that communities are decision-makers on how land should be used.

The ongoing commercialisation of airports driven by a host of neoliberal factors has resulted in the expansion of airport use for commercial purposes unrelated to aviation. This trend has emerged because of the agglomeration offered by airports in accessing regional, national, and international markets (Schaafsma et al. 2008). Agglomeration effects offered by airports are responsible for breaking down state-run national aviation systems into smaller, independent business units that function in a new mixed land use system used for aviation and commercial purposes (Freestone & Wiesel 2014; Freestone 2011). The transition from simple airport spaces to the aerotropolis model of development requires a change in land use management. Territorial contestations of airport land transformation often complement this change, as is evident in the case of Essendon Airport, Australia. Freestone and Wiesel (2014: 280) contend that the airport is a classic case of transforming a 'run-down airport general aviation facility into an emergent business and retail hub in Melbourne's north-western suburbs'. The mixed use of the airport has enhanced the town's economic activity and demonstrated some of the prospects associated with the commercialisation of airports. Importantly, this commercialisation has potential employment benefits as well.

Contestations in airport land use emerge from different interests of diverse stakeholders in the continued expansion of aeronautical and non-aeronautical functions in the making of airport cities. In India, contestations over airport land emerged because of the rushed and unplanned territorial expansion of the aeronautical and non-aeronautical commercial functions of airports. Jain and Watve (2019) argue that contestations unravel because airports are built without the consent of local communities and without the consideration of the environmental impact. The lack of consultation and environmental concern hampers sustainable livelihood strategies deployed by the people close to the Greenfield Airport, Mopa. These hastened developments are underpinned by political motivations for re-election and create livelihood vulnerabilities for communities that contest the expansion of these developments (Jain & Watve 2019). This is because the displacement and dispossession of these communities undermine their human and socioeconomic rights in the face of neoliberal expansions. In the case of Durban, this is evident in how unhappy businessmen and the political elite use community protests to stop megaprojects.

Megaprojects were established to improve a territory's economic development infrastructure and maximise economic growth (Broudehoux 2017). However, the pursuit of economic growth through major projects produces different outcomes because of the contested costs and benefits of such projects. In the proposed expansion of the Durban harbour, contestation emerged on the environmental impact of the expansion on the basis that it undermined the Clairwood Racecourse, which consists of a major wetland area in the city, serving as a habitat for birds, amphibians, and other fauna and flora (Mpungose 2017). In the province of Gauteng, South Africa, the Gautrain project, completed in 2013, is criticised for deepening mobility-related exclusion and prioritising the wealthy instead of the poor. This is because of high transportation costs, whereas public funds and an integrated transportation plan were used, which failed to cater to low-income travellers (Thomas 2013). This can be linked to the criticism by Mpungos (2017) that megaprojects only address the needs of the elite and increase inequality, marked by cost overruns, and neglect the adverse effects of this notion.

In Durban, the benefits of large-scale projects remain a space of contestation and debate by local politicians, city officials, research institutions, and civil society organisations because of the diverse interests and power relations among the primary beneficiaries of megaprojects (Martel & Sutherland 2018). Megaprojects are also criticised for not producing the required substantial socioeconomic benefits and sustainable impact on the lives of ordinary South Africans (Robbins 2015). Such projects create contrasting economic realities in the city of Durban and contribute to the contested benefits, costs, and risks in urban and rural settings (Sutherland et al. 2016). While many of the dispossessions for megaprojects are proposed for social and economic development and job opportunities, Doucet (2013) argued that the use of public money for megaprojects does not create the promised investment and job opportunities. The basis of some of these projects on trickle-down economics results in dispossessed and displaced groups being marginalised in economic opportunities, who often have to wait for years to experience tangible socioeconomic benefits.

The Durban Aerotropolis in its territorial context

The Durban Aerotropolis is situated at King Shaka International Airport (KSIA), established on 1 May 2010, and it is located in the northern part of Kwa-Zulu-Natal Province, South Africa. The airport has 38 facilities comprising banks, food outlets, clothing stores, and other retail services (Airports Company South Africa 2021). The building of the airport, which cost R79 billion, 'boasts [of] a passenger terminal building floor area of 103,000 m^{2}' (NAG 2015). KSIA serves as the gateway to the province, which is widely known for its rich history and tourism sector. The birth of the Durban Aerotropolis is regarded as a significant milestone in establishing Durban as a trade and business hub in Africa, because the city also acts as an entrepôt, a broker, between the country's seaport corridor and landlocked provinces, such as Gauteng, Free State, Mpumalanga, and landlocked countries such as Swaziland, Botswana, and Lesotho. Through the relational use of the city's seaport and the freight airport, the Durban Aerotropolis is centred on freight logistics strongly linked to the seaport, with the capacity of handling 31.4 million tonnes of cargo annually and a container capacity of 2.9 million (Dube Tradeport 2019). The emergence of KSIA presents an opportunity for Durban to benefit more from the seaport and the emerging aerotropolis as drivers of local economic activities through investment, employment, and infrastructure development.

Urbanisation trends associated with the aerotropolis are increasingly observable through investments, such as the Tongaat Hulett Property Development for residential and infrastructure development and the recent Cornubia Industrial and Business Estate. These investments contribute to the Durban Aerotropolis because the latter is within the airport's spatial corridor. These trends follow an international practice in which the initial stages of aerotropolis development are supported by investment in real estate, shopping complexes, and industrial complexes that make up the economic base connected to aeronautical functions. Australia has a classic example of this practice, where a dilapidating airport town was supported by establishing retail and business centres linked to the airport and the needs of local citizens (Freestone & Wiesel 2014). In the case of KSIA, rapid development and investment along the airport route can be understood as purposefully controlled investments to build infrastructure to support the emerging aerotropolis. After all, large manufacturing and logistical industries are made up of smaller business units that depend on the agglomeration effects of the connectedness of emerging businesses.

The Durban Aerotropolis is significant because it also offers road connectivity to landlocked territories and countries. The aerotropolis was established to strengthen regional economic development using its nodal route (Department of Economic Development, Tourism and Environmental Affairs 2015). At the heart of the aerotropolis is the need to transform the province's economy through employment creation, because KwaZulu-Natal is one of the most impoverished provinces in South Africa. The province constitutes approximately 20% of the South African population; it has a 34% poverty rate using the food poverty line and 52% when using the lower-bound poverty line, with these statistics increasing to 68% when using the upper-bound poverty line (Statistics South Africa 2020). The Dube Tradeport SEZ has been playing a leading role in advancing the linkage of KSIA with export-oriented manufacturers and agricultural zones and in facilitating foreign investment. It has also been playing a significant role in the integrated regional spatial planning and implementation of the Durban Aerotropolis, and it provides the extra-connectivity infrastructure required to export goods such as petroleum, fuel, chemical and rubber products, clothing, textiles, and metals from the manufacturing industry located in the city centre (Dube Tradeport 2019).

While the Durban Aerotropolis initiative is strategically oriented towards increasing the competitiveness of Durban and the entire province, the KSIA airport is positioned between the iLembe District and the Durban Metropolitan area. Figure 7.1 shows the map of KwaZulu-Natal and the proximity of both regions to the Durban Aerotropolis.

Both areas, particularly the iLembe District, are famous because of their rich Zulu Kingdom heritage (iLembe District Municipality 2019). Furthermore, both districts are situated within the 35 km radius of the aerotropolis. However, such shared territorial benefits and costs between a city, a small town, and a rural area have not been thoroughly studied in the context of an aerotropolis. As observed by Prosperi (2007), polycentric metropolitan morphologies emerge from new centres

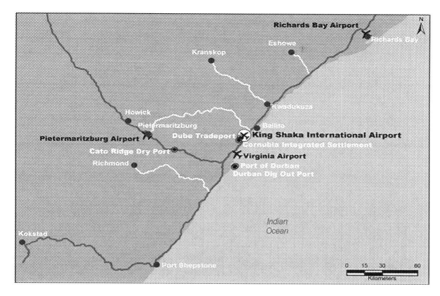

Figure 7.1 Durban Aerotropolis and the areas within one-hour radius.
Source: Luthuli (2018).

of regional growth that feed into the broader regional economic development landscape of the metropolis. Balancing the benefits and costs of these new economic centres is important for ensuring spatial rebalancing, which redirects investment to lagging regions to catch up with economically better-developed regions.

Development contestations occur in various ways. Sutherland et al. (2018: 355) observed that contestations and collaboration between the state, citizens, and the private sector over large-scale economic development projects and dual governance systems are constant features in Durban. Safitri (2014) held that institutions are spaces for the social interactions, negotiations, and contestations of diverse actors and interest groups. The case of the Durban Aerotropolis is significant because it diverts the focus from contestations between citizens and developers to contestations that happen between territorial institutions in the form of local governments pursuing economic advantages from shared development opportunities. In this context, both the eThekwini Municipality and the iLembe District utilise tourism services and manufacturing as strategic economic drivers (eThekwini Municipality 2020), thereby creating economic competition. Pursuing these economic benefits in shared territories can act as underlying motives for contestations within different institutional settings.

A comparative overview of the two municipalities shows the social and economic distinctiveness of these areas. The iLembe municipality is classified as a district municipality by virtue of being made up of four other local municipalities, whereas eThekwini is a metropolitan municipality.

Table 7.1 Comparison of eThekwini and iLembe

Segments	eThekwini	iLembe
Population	3,987,648	657,612
Unemployment	22%	30.6%
Economic sectors	Manufacturing, finance, wholesale, transport, tourism, and government services	Tourism, agriculture, construction, and government services
Budget	R52.3 billion	R1.2 billion

Source: Author based on eThekwini IDP (2021); iLembe District Municipality (2020).

As seen in Table 7.1, the two municipalities have distinct socioeconomic profiles that complement the demographic information of the two territories. EThekwini, by virtue of being a metropolitan municipality, has a higher population rate than the iLembe District. This is because cities are centres of economic growth and job creation, attracting rural–urban migration. The economic sectors between the two territories are diversified, with iLembe District's economy concentrated on agriculture, tourism, and government services after years of decline in manufacturing sites, such as Isithebe, Mandeni. In contrast, eThekwini is largely driven by the industrial economy. The disparity in the budget between the two territories also reveals the uneven economic power between these municipalities. These differences are likely to determine the bargaining power of these municipalities when attracting investment for the aerotropolis and influence economic development outcomes. Bargaining power also has implications for spatial disparities between city and rural areas.

Economic cooperation: emerging and underlying territorial contestations

The Durban Aerotropolis strategically connects to the eThekwini Municipality's Strategic Infrastructure Projects (SIPs) in the Durban–Gauteng Corridor, which 'forms the backbone of South Africa's freight transportation network and will improve the countries' ability to raise transport efficiencies, unlock the development potential of various areas and improve freight handling capacities to meet future demands' (eThekwini Municipality 2020). However, it is also strategically linked to the Durban–iLembe and Mozambique Corridor that promotes the integration of infrastructure between eThekwini Municipality and iLembe District. From a space-neutral approach, eThekwini Municipality promotes the integration of small-town and rural municipalities, which are isolated from the leading logistical platforms, into surrounding corridors (eThekwini Municipality 2020). This integration is more important for eThekwini and iLembe, based on the central location of the aerotropolis, thereby pointing to the need for the municipalities to cooperate in development initiatives along the aerotropolis corridor. The cooperation of both local governments is instrumental in improving institutional arrangements for infrastructure development and economic integration.

Promoting mutual economic benefits through the Durban Aerotropolis is essential, as it aligns with the country's district development model (DDM), which seeks to ensure enhanced cooperation and prevent municipalities from working in silos. The DDM aims to address the culture of working in silos between municipal and government entities at the regional level. It promotes cooperative governance in development planning (Department of Planning, Monitoring and Evaluation 2019). In line with the DDM approach, the eThekwini municipal government initiated a series of bilateral meetings with neighbouring municipalities to address cohesion issues between the spatial development framework (SDF)[2] and the Aerotropolis Master Plan linked to the N3 Corridor[3] (eThekwini Municipality 2020). The Mandeni and KwaDukuza Municipalities, under iLembe District, were thoroughly consulted to ensure enhanced cooperation in the SDF (iLembe District Municipality 2021). This process is essential because it reveals the need to consult neighbouring regions in developing SDFs to ensure the economic integration of the city and neighbouring small towns and municipalities. Todes and Turok (2018) also argued for cities and national structures to create policies that enable neighbouring small-town and rural municipalities to be economically integrated into the benefits emerging from cities.

Although economic geography theories argue that lagging regions should be economically integrated into cities, such as eThekwini Municipality in this context, it is also important for small-town and rural municipalities in iLembe District to use their strategic locations to benefit from the Durban Aerotropolis. As the aerotropolis is centrally located between eThekwini and iLembe, it prompts both municipalities to use their competitive advantages to benefit economically from it. Place-based economic development approaches argue that each locality uses its inherent resources, stakeholders, and strategic advantages to tap into its economic potential (Barca et al. 2012). ILembe District, which is strategically placed between two major ports, the Durban and Richards Bay and the N3, utilises place-based economic development strategies (iLembe District Municipality 2021). However, eThekwini utilises a space-neutral approach that seeks to keep the investment within the city to maximise its economic potential. This reveals the underlying contestations utilised by the two municipal governments based on conflicting economic geography and development planning strategies. However, it is also essential to understand which areas of contestation need to be examined by both municipalities to ensure minimal territorial economic contestations in the emerging aerotropolis.

ILembe District has four municipalities under its geographical scope: KwaDukuza and Mandeni are small-town municipalities, whereas Ndwedwe and Maphumulo are rural municipalities. The iLembe District SDF notes that KwaDukuza and Mandeni comprise urbanisation and industrial development underpinned by connectivity to KSIA and aerotropolis development (iLembe District Municipality 2021). Both municipalities seek to benefit from the aerotropolis development through economic nodes that converge at the Durban Aerotropolis, with development in iLembe emerging from the north to the south and that of eThekwini emerging from the south to the north. Furthermore, both municipalities seek to economically benefit from the extra connectivity offered by KSIA and the Dube Tradeport SEZ, but they create underlying contestations instead by utilising economic development strategies that are antagonised by pursuing similar economic benefits from opposing ends. These problems, particularly space-neutral economic development approaches, have been argued to cause regional divergence in the past two decades (Altman & Robin 2018).

The utilisation of differing economic development approaches (space-neutral and place-based) is the basis of the emerging territorial contestations that play out in the development landscape. The industrial hubs in KwaDukuza and Mandeni can be expanded through further investment in the aerotropolis corridor to promote export-oriented manufacturing through a place-based economic development approach. The Aerotropolis Master Plan also identifies the need for these municipalities to design SDF plans aligned with the Durban Aerotropolis (Dube Tradeport 2017). However, such plans ultimately stem from each locality seeking its competitive advantage, with non-metropolitan municipalities likely to lag behind because of the agglomeration effects offered by cities. Manufacturing in KwaDukuza and Mandeni can be extended to support aeronautical functions by establishing aviation-related construction factories at the Dube Tradeport SEZ. This could support the commercial or non-aeronautical functions of the airport. However, since the emergence of the Durban Aerotropolis Masterplan, strategic initiatives in tourism and industrial and residential infrastructure, such as Sibaya Precinct Nodes 1 and 5, have shown the evident development of the Durban Aerotropolis link to the Durban city centre. These developments prove to be pro-cities, because investments tend to flow into cities rather than into small-town and rural municipalities.

Many areas of territorial economic contestation between iLembe and eThekwini municipalities are evident based on several economic drivers in the region. Both territories benefit from the connectivity offered by KSIA, particularly in the tourism sector (DEDTEA 2015). Based on being coastal geographies, both territories compete for tourism linked to the rich history of the Zulu Kingdom as heritage sites and the affluent areas of Zimbali Coast on iLembe District by not exploring the shared benefits of tourism. Tourism makes up 8% of the GDP of eThekwini and 15.6% of that of iLembe (Durban EDGE 2021; Enterprise iLembe Development Agency 2019). In competing as tourism destination sites, both territories have seen significant investment in tourism and residential accommodation-related tourism (iLembe District Municipality 2020; eThekwini Municipality 2020). This reflects the importance of tourism in South Africa's economic development landscape. Rogerson (2016) observed that tourism has been one of the success stories in South Africa's local economic development initiatives. Essentially, tourism-related activities, such as heritage sites, wildlife viewing, nature trails, birding, and coastal beaches, rely on the connectivity offered by KSIA and add economic value to the emerging aerotropolis. Tourism has also been a pillar for development in iLembe, as envisaged as one of the key strategic focus areas for the Enterprise iLembe Development Agency (2020). The shared benefits of tourism can be maximised by creating linkages between tourism activities in the two municipalities.

Territorial contestations and development prospects

Territorial contestations pose significant barriers to investment and regional economic development. The formal and informal processes in constructing regional economic nodes consist of engagement and negotiation between different stakeholders. So far, engagements in the Durban Aerotropolis have been based on ensuring trickle benefits to neighbouring municipalities instead of seeing these municipalities as contributors to aerotropolis construction and development. Within emerging contestations, 'vulnerability to conflict in peri-urban areas can be attributed to the interaction of macro-level processes with local-level factors, including diverse claims, overlapping legal and governance frameworks and, critically, local power relations' (Lombard 2016: 2700). The focus on territorial institutions in the form of local governments pursuing economic developmental outcomes for their respective localities reflects a new angle in the debate on contested urban land. The case study of the Durban Aerotropolis demonstrates the importance of local governments in setting up institutional arrangements underpinned using an airport as a regional broker. However, the problem is the competitive nature of economic development, as territories compete for investment and exploit business opportunities, thereby leading to municipalities working in silos. Through KSIA as a broker, the power relations between the two municipalities create contrasting realities that require cooperative governance in managing and sharing costs and benefits to promote equitable development and minimise spatial disparities.

A significant focus of the Durban Aerotropolis is on the connectivity offered by KSIA and the immediate benefits of the emerging aerotropolis. Luthuli and Houghton (2019) focus on the connectivity of the aerotropolis to nodes, such as Howick, Pietermaritzburg, Kranskop, Eshowe, Mandeni, and Port Shepstone. However, these nodes need to be extended to inland areas, such as Ndwedwe, because of their proximity to the benefits of employment and infrastructure development offered by the aerotropolis. The significance of the brokerage role played by the aerotropolis is important, but so is focusing on the contribution to local development and contestations that may underpin such development. Given that both municipalities utilise different economic development strategies that do not reconcile development (place-based vs. space-neutral), the developmental goals pursued by either municipality may threaten equitable development if economic benefits are not shared equally. This is already evident in the concentration of investment in Durban – because investment is biased towards pro-urban areas. These trends cause iLembe's small-town and rural municipalities to wait for the trickle-down of investment and economic benefits, forcing the region to catch up with developed cities. This phenomenon is cited as the main driver of regional inequalities across many countries.

Investment needs to be diverted to iLembe District to maximise the district's economic potential. Currently, iLembe's SDF (2021: 72) underscores that the district will benefit 'as an integral support district within the Aerotropolis Master Plan through skills transfer, housing provision to sustain demand/need and internal growth of local economic nodes and entrepreneurial enterprises'. This pits the district into playing a secondary role by supporting the aerotropolis instead of immersing the district in the aerotropolis. Essentially, this promotes a space-neutral position in which small towns and rural municipalities must link their development plans to economically prosperous neighbouring cities and benefit from trickle-down. Todes and Turok (2018) emphasised the need to connect lagging regions to cities in South Africa because of the increasing decline in small-town and rural municipalities in recent years. However, evidence from the United States shows that such approaches have not worked over the last few decades and have only furthered spatial inequalities (National League of Councils 2019). This approach is problematic in the South African context because it promotes dependence on big cities, which stifles investment in rural and small-town municipalities and the ability to generate innovative economic activities. This has resulted in the concentration of place-based development strategies through using a locality's economic potential as the driver for development.

Territorial conflicts are enhanced by the underpinning economic development approaches applied by different actors to pursue similar developmental outcomes. The free movement of people between the two regions is important for sharing employment opportunities, but the connectivity and brokerage role of firms and the creation of employment prospects also need to be shared based on each locality's needs. Given the lack of inward investment for the development of business entities that will support the aeronautical and non-aeronautical needs of the airport, iLembe District must look into inward investment and tap into existing resources and stakeholders in identifying valuable opportunities that link to the immediate needs of the aerotropolis. These needs should not be limited to hospitality tourism and housing estates only, but they should also comprise business entities that are the basis of economic development. The coordinated clustering of business entities is one of the success factors in aerotropolis development (Liou et al. 2016).

Luthuli and Houghton (2019: 203) observed that 'the greenfields nature of the aerotropolis also means that it is a space incorporating a multitude of natural habitats which need protection'. This signifies the importance of conserving greenfields when undertaking development and also ensuring that greenfields can be incorporated into new development opportunities. Such an approach is evident in expanding the agrizone in the Dube Tradeport area and developing eco-friendly facilities to promote a range of green initiatives (Dube Tradeport 2021). This approach is distinctive from traditional industrial development approaches based on destroying greenfields in pursuit of capital interest and expansion. A notable example of this development approach is the contested expansion of the Durban port harbour because of its devastating effects on greenfields and other natural habitats linked to biodiversity. Therefore, the case of the Durban Aerotropolis is important because it contributes to the province's greenfields sustainably while also battling territorial contestations for economic development opportunities.

The development of the Durban Aerotropolis can create economic convergence between eThekwini and iLembe Municipalities because both localities are embarking on development linked to the aerotropolis. This signals the brokerage role played by the Durban Aerotropolis because it cuts across various municipal boundaries, promoting development that knows no boundaries, despite the spatial administrative boundaries. While this is an important contribution to the development landscape, future developments are likely to require a multi-level governance approach, because Dube Aerotropolis serves multiple corridors and municipal boundaries. However, the fundamental problem is that these developments have primarily focused on hospitality, tourism, and residential opportunities to attract tourists, which hides the underlying economic contestations. It is important for these territories to ensure cooperative governance and intergovernmental relations in line with the DDM to enhance levels of cooperation in the making of the Durban Aerotropolis.

Conclusion

The continued globalisation and trade needs have consistently shaped the need for expedient connectivity in global value chains and the movement of people. In South Africa, these needs have been met by the novel and the evolving role played by airports and aviation industry in meeting the high demand for transportation of people, goods, and services. Evidence shows that the construction of megaprojects like new airports is not without contestations, as is evident in Australia, Indonesia, India, and Kenya. Contestations cannot be avoided when people are displaced from their land for economic reasons. While contestations are mainly studied in relation to communities and developers, the study of the South African context is novel because it adopts an institutionalist view by focusing on territorial economic contestations that underpin the Durban Aerotropolis.

Based on the central location of the aerotropolis between eThekwini and iLembe Districts, the chapter delineated different economic development strategies. While urban eThekwini utilises a space-neutral approach that believes neighbouring small-town and rural municipalities should be plugged into its agglomeration and benefit from trickle-down processes, this approach undermines the diversion of investment to rural iLembe. This is noted in several investments occurring in eThekwini Municipal District – heading to the KSIA, with a slower pace of investment from iLembe District. The district's SDF acknowledges economic benefits from the aerotropolis but defines the role of iLembe as a support system of the aerotropolis instead of tapping into the existing economic potential from a place-based economic development strategy. It requires a change in political thinking to ensure that the district takes full advantage of the economic potential offered by the aerotropolis.

Tourism and industrial zones are perceived and used by both local governments as employment generators. Given the need to ensure that export-oriented manufacturing and business entities support the aeronautical and non-aeronautical functions of the aerotropolis, both regions contest for investment that will address this emerging problem. Currently, only hospitality and tourism needs are catered to in the Durban Aerotropolis. This is through the development of a series of residential estates that form part of the highly sought tourism in industries offered by the territories, while there is neglect of the industrial component of aerotropolis development. Key lessons emerge from the need for these territories to strengthen institutional arrangements to tap into the shared benefits of the aerotropolis to ensure balanced development and growth. At the heart of this process is the need to ensure that investment does not flow to cities only because they offer attractive agglomeration effects, but the government ensures that lagging regions should also benefit from investments that promote their regional competitiveness.

Notes

1 South Africa has three levels of government: national, provincial, and local government. The local government system has three categories: local municipalities, district municipalities, and metropolitan municipalities.
2 The spatial development framework is a municipal policy document that represents the spatial distribution of land use.
3 The N3 Corridor is the national highway that links KwaZulu-Natal with other provinces and landlocked countries.

References

Airports Company South Africa (2021). *King Shaka International Airport*. Durban: South Africa.
Altman, R. C. and R. E. Rubin (2018). 'Foreword'. In J. Shambaug and R. Nunn (eds.), *Place-based policies for shared economic growth*. Washington, DC: The Brookings Institution, 1–4.
Banai, R. (2017). 'The aerotropolis: Urban sustainability perspectives from the regional city'. *Journal of Transport and Land Use* 10(1), 357–373.
Barca, F., P. McCann, and A. Rodriguez-Pose (2012). 'The case for regional development intervention: Place-based versus place-neutral approaches'. *Journal of Regional Science* 52(1), 134–152.
Bond, P. (2015). 'Durban's port-petrochemical complex as a site of economic and environmental violence'. *Rethinking Cities in the Global South: Urban Violence, Social Inequality and Spatial Justice*. Tata Institute, Mumbai, 19 January.
Borras, S. M., Jr. and J. C. Franco (2012). 'Global land grabbing and trajectories of agrarian change'. *Journal of Agrarian Change* 12(1), 34–59.
Bridger, R. (2017). 'Airport-linked special economic zones, aerotropolis projects and the race to the bottom'. *Global Tax Justice at Crossroads*. London, 5–7 July.
Broudehoux, A. (2017). *Mega-events and urban image construction*. London: Routledge.
Button, K. (2010). 'Economic aspects of regional airport development'. *WIT Transactions on State of the Art in Science and Engineering* 38, 9–25.
Charles, M. B., P. Barnes, N. Ryan, and J. Clayton (2007). 'Airport futures: Towards a critique of the aerotropolis model'. *Futures* 39(9), 1009–1028.
Da Silva, S., K. B. Nielsen, and H. P. Bedi (2020). 'Land use planning, dispossession and contestation in Goa, India'. *The Journal of Peasant Studies* 47(6), 1301–1326.
Department of Economic Development, Tourism and Environmental Affairs (DEDTEA) (2015). *KwaZulu-Natal integrated KZN integrated aerotropolis strategy*. Pietermaritzburg: KwaZulu-Natal.
Department of Planning, Monitoring and Evaluation (2019). *Khawuleza district development model*. Pretoria: DPME.
Doucet, B. (2013). 'Variations of the entrepreneurial city: Goals, roles and visions.' *International Journal of Urban and Regional Research* 37(6), 2035–2051.
Dube Tradeport (2017). *Durban aerotropolis masterplan*. Durban: Dube Tradeport.

Dube Tradeport (2019). *Annual report 2019*. Durban: Dube Tradeport.
Dube Tradeport (2021). *Annual report 2020–21*. Durban: Dube Tradeport.
Durban EDGE (2021). *eThekwini economy*. Durban: Durban EDGE.
Economic Development, Tourism and Environmental Affairs (2015). *KwaZulu-Natal integrated KZN integrated aerotropolis strategy*. Pietermaritzburg: EDTEA.
Enterprise iLembe Development Agency (2019). *Annual report 2018–19*. Ballito: Enterprise iLembe Development Agency.
Enterprise iLembe Development Agency (2020). *Annual report 2019–20*. Ballito: Enterprise iLembe Development Agency.
eThekwini Municipality (2020). *eThekwini municipality 2020/2021 IDP*. Durban: eThekwini Municipality.
eThekwini Municipality (2021). *Integrated Development Plan 2021/22*. Durban: eThekwini Municipality.
Freestone, R. (2011).' Managing neoliberal urban spaces: Commercial property development at Australian airports'. *Geographical Research* 49(2), 115–131.
Freestone, R., and D. Baker (2011). 'Spatial planning models of airport-driven urban development'. *Journal of Planning Literature* 26(3), 263–279.
Freestone, R. and I. Wiesel (2014). 'The making of an Australian 'airport city''. *Geographical Research* 52(3), 280–295.
Goetz, A. R. and B. Graham (2004). 'Air transport globalisation, liberalisation and sustainability: Post-2001 dynamics in the United States and Europe'. *Journal of Transport Geography* 12(4), 265–276.
Huka, F. (2021). 'Rights violations in Isiolo international airport land expropriation'. Retrieved 05.06.2021. www.theelephant.info/op-eds/2021/06/04/rights-violations-in-isiolo-international-airport-land-expropriation/
iLembe District Municipality (2019). *iLembe District Municipality integrated development plan*. KwaDukuza: iLembe District Municipality.
iLembe District Municipality (2020). *Integrated development plan 2021/2022*. KwaDukuza: iLembe District Municipality.
iLembe District Municipality (2021). *Spatial development framework*. KwaDukuza: iLembe District Municipality.
Jain, N. and A. Watve (2019). 'Contestation of environmental impact assessment for Greenfield Airport, Mopa, India through the lens of livelihood vulnerability'. *International Journal of Scientific and Engineering Research* 10(12), 314–322.
Kasarda, J. D. (2006). 'Logistics and the rise of the aerotropolis'. *Real Estate Issues* 25, 43–48.
Kasarda, J. D. (2011). 'Creating an aerotropolis: How Indianapolis is strategically charting its Airport's and region's future'. *Global Airport Cities* 5(1), 16–18.
Kasarda, J. D. (2013). 'Aerotropolis: Business mobility and urban competitiveness in the 21st century'. In *Culture and mobility*. K. Benesch (ed.). Bonn: Universitätsverlag, 9–20.
Liou, J. J., C. C. Hsu, and Y. C. Chaung (2016). 'Using a hybrid model to explore the key factors for a successful aerotropolis'. *5th International Congress on Advanced Applied Informatics*. 10–14 July, Kumamoto, Japan.
Lombard, M. (2016). 'Land conflict in peri-urban areas: Exploring the effects of land reform on informal settlement in Mexico'. *Urban Studies* 53(13), 2700–2720.
Luthuli, N. H. (2018). *An exploration of the conceptualisation and enactment of regional economic development through an analysis of the Durban aerotropolis in KwaZulu-Natal, South Africa*. Doctoral dissertation. University of Kwa-Zulu Natal, Durban.
Luthuli, N. and J. Houghton (2019). 'Towards regional economic development in South Africa: Conceptualising the 'region' associated with economic development through the Durban Aerotropolis'. *Urbani Izziv* 30, 194–211.

Martel, P. and C. Sutherland (2018). 'Durban's back of port project: A local spatial knowledge production process framed by urban entrepreneurialism'. *Urban Forum* 29(4), 397–412.

McDonough, E. (2014). *Global flows, local conflicts and the challenge of urban governance: Managing the city-airport interface in London, UK.* The Spaces and Scales of Global Flows. London: International Airports and Globalised Urbanisation.

Mpungose, A. (2017). *The politics of megaprojects: Assessing the socio-spatial and environmental impacts of the proposed dig-out port in south Durban.* Doctoral dissertation. Durban: University of KwaZulu-Natal.

National League of Cities (2019). *Place-Based Policies for America's Innovation Economy.* New York: National League of Cities.

Netherlands Enterprise Agency (NAG) (2015). *Fact finding southern Africa.* Zoetermeer: NAG.

Prosperi, D. C. (2007). 'Airports as centers of economic activity: Empirical evidence from three US metropolitan areas'. *REAL CORP 007 Proceedings*, Tagungsband Vienna, 20–23 May.

Robbins, G. (2015). 'The Dube TradePort-King Shaka International Airport mega-project: Exploring impacts in the context of multi-scalar governance processes'. *Habitat International* 45(3), 196–204.

Rogerson, C. M. (2016). 'Urban tourism, aerotropolis and local economic development planning: Ekurhuleni and O.R. Tambo International Airport, South Africa'. *Miscellanea Geographica* 22(3), 1–7.

Safitri (2014). *Conflict, development and livelihood transition: A case study of the international airport development in Lombok Nusa Tenggara Barat Indonesia.* Master of Science dissertation. Wageningen University, Wageningen.

Schaafsma, M., J. Amkreutz, and M. Guller (2008). *Airport and city: Airport corridors: Drivers of economic development.* Amsterdam: Schiphol Real Estate.

Sigler, T. J. (2013). 'Relational cities: Doha, Panama City, and Dubai as 21st century entrepots'. *Urban Geography* 34(5), 612–633.

Statistics South Africa (2020). *Quantity labour force Q3.* Pretoria: Statistics South Africa.

Sutherland, C., D. Khumalo, V. Sim, and S. Buthelezi (2016). 'Social constructions of environmental services in a rapidly densifying peri-urban area under dual governance in Durban, South Africa'. *Bothalia-African Biodiversity & Conservation* 46(2), 1–18.

Sutherland, C., D. Scott, E. Nel, and A. Nel (2018). 'Conceptualizing "the urban" through the lens of Durban, South Africa'. *Urban Forum* 29(4), 333–350.

Thomas, D. P. (2013). 'The Gautrain project in South Africa: A cautionary tale'. *Journal of Contemporary African Studies* 31(1), 77–94.

Todes, A. and I. Turok (2018). 'The role of place-based policies in economic development'. *Progress in Planning* 123, 1–31.

8 Competing aspirations and contestations at Isiolo International Airport, Kenya

*Evelyne Atieno Owino and
Clifford Collins Omondi Okwany*

Introduction: Isiolo, a failed but conflict-stirring airport

The large-scale infrastructure of Isiolo International Airport has created contestation between the developers and nomadic pastoral communities. While these nomadic pastoralists (Borana, Turkana, Somali, and Samburu) are the principal inhabitants of Isiolo, some agriculturalists – the Meru ethnic communities – also live in the region. The airport project created controversy between these ethnic groups over land, giving rise to economic anticipation, land speculation, and alienation – selling of community land to private entities; this is despite the fact that land in Isiolo County is not yet registered as community land but still under the trust land owned by the national government. It led to competition between actors when the Kenyan government communicated its intention to upgrade the airport from an airstrip to an international airport in 2004. The Kenyan government believed that this upgrade became necessary because transporting meat products from the newly built abattoir in Isiolo to markets such as the Middle East needed to become quicker and easier. The government also planned to permit the transport of the khat plant (*miraa catha edulis*) to international markets, such as Somalia and the United Kingdom, and to reduce traffic congestion at the Wilson Airport in Nairobi (Owino 2019). However, the land allocation to the affected communities due to the airport's expansion has created grievances between the agricultural Meru communities and the nomadic pastoral Borana communities in Isiolo. This has lately been seen as a justification for the emergence of radical groups, such as community-based armed groups and violent extremist organisations (for example, Hansen et al. 2019; Okwany 2016, 2020a, 2023).

The contestation and competition over land are due to the government's neglect of the north. In some cases, the government upholds security laws in case of emergency (Okwany 2023; Okwany et al. 2023). We later demonstrate how the Kenyan government at times uplifts the law but also disregards it when it has insufficient capacity or follows certain state security advantages. In addition, the international airport promised economic benefits, such as an increase in tourism, an easier reach of markets for other products, and a reduction of road accidents. Isiolo airport is a critical case to study because it exemplifies the impact of development dynamics and conflicts and shows how the future of Isiolo may look like with the rapid

DOI: 10.4324/9781003494966-8

This chapter has been made available under a CC-BY-SA 4.0 license.

socioeconomic transformations taking place. Its expansion triggered contestation of property rights and landownership (Mkutu and Boru 2019). Isiolo is one of the most intense areas faced with speculation and anticipation; people predict, sell, and buy property, hoping for value addition of land due to the emerging large-scale or mega-infrastructure projects in Kenya (Enns 2019).

The upgraded Isiolo International Airport is built on an 815 ac site, and it is 283 km away from Nairobi City. It was established on disputed land between Wabera ward in Isiolo County and the Nyambene area bordering Tigania East and West constituencies in Meru County. Planned in three phases, the construction of the airport began in 2011, and in its completion, the Kenyan government spent USD12.82 million or 2.7 billion Kenya shillings on its renovation to an international standard airport (07.05.2019, *Business Daily*). It was commissioned to be part of the Lamu Port–South Sudan–Ethiopia Transport (LAPSSET) corridor development project and the Isiolo export-oriented abattoir.

The abattoir is envisioned to process an estimated 474,000 animals annually, a means of boosting and benefiting about 200,000 pastoral livelihoods, and it received 800 million Kenya shillings from the World Bank (19.07.2020 *Daily Nation*), yet after 17 years, the slaughterhouse is yet to be operational. Both the LAPSSET and the abattoir were envisioned in 2007, and the development of the abattoir began the same year. The airport is the fifth international airport in Kenya, joining Jomo Kenyatta, Moi International Airport in Mombasa, Kisumu, and Eldoret International Airports, and it is operated by the Kenya Airport Authority (KAA), a government-owned enterprise, and it was intended to handle a capacity of 125,000 passengers annually (Airport Technology 2017).

The airport was to boost the economy of the northern region, serving counties such as Isiolo, Marsabit, Meru, Laikipia, and part of Samburu. Although the airport was completed in 2017, two years later, some of its units were closed down, with certain aircraft companies citing the discontinuation of services due to a lack of passengers.[1] Transportation of flowers, khat, and meat was considered cheaper via road than by air.[2] However, the government maintained its rhetoric, contradicting the earlier claims of the 2017 completion, pointing out that the completion by September 2021 aims to boost the airport's usefulness.[3] Yet at the time of writing, the airport still had not re-opened the closed units; part of it is operating, but not as the earlier-envisioned international-standard airport.

Empirical research on the competing aspirations and contestations accompanying mega-infrastructure projects in pastoral rangelands is limited, leaving gaps in understanding their implications for the northern frontier. Here, it is important to note that despite large infrastructures such as Isiolo International Airport, the abattoir, the Lamu Port–South Sudan–Ethiopia Transport (LAPSSET, see later text) corridor, or the Crocodile Jaw Dam project, most of the land in Isiolo County remains unregistered. This study delves into the dynamics of conflict arising from infrastructural projects. Focusing on the pastoral corridor, we use empirical evidence to demonstrate the varying aspirations and contestations among stakeholders following the airport's expansion.

We conducted 36 qualitative interviews and eight focus group discussions (FGDs) that took place between 2017 and 2019 and follow-up discussions in 2020 and 2021, including a conference on the implementation of the Community Land Act in June 2022.[4] The interviews involved key informants, such as government security officers, local politicians such as members of the County Assembly, land registrars and adjudicators, ward administrators, court officers, as well as community members.

Our chapter begins with a brief history of the Northern Frontier District, ethnicity, and politics in Isiolo. Thereafter, we relate the concept of the frontier and the land tenure system before providing empirical evidence on how the airport-induced displacements, irregularities regarding compensation, challenges of land re-allocation, and local political struggle over land. Lastly, we demonstrate that Isiolo International Airport risks being an economically redundant project. We triangulated the interviews on the progress of LAPSSET, the airport land contestations, competitions, and the dynamics of violence in Isiolo with academic sources, showing the local perspectives on the airport and conflicts over land in Wabera and Ngaremara wards.

History of the Northern Frontier District, ethnicity, and politics in Isiolo

Arid areas in northern Kenya have been treated as unworthy land since the colonial British East Africa, for they could not serve agricultural production. These areas, also referred to as the northern frontier, were thus politically neglected by the colonial and national governments[5] and perceived as what Elliott (2016: 512) refers to as 'Kenya B'. Recently, they have become a hub for infrastructure developments, such as the airport, the highway, and the railway under LAPSSET.

Intra- and inter-pastoral conflicts over pasture and water, and cattle raiding, increased in Isiolo County. It also has become a hub for illegal arms trade (Mkutu 2019). Our field research from 2017 to 2021 confirms that banditry, cattle rustling, border grazing disputes, and community armed groups were the main contributors to conflict in Isiolo. Despite ongoing land disputes since the 1990s, infrastructural projects along the pastoral corridor are bringing development, making Isiolo the gateway link between the north and the southern parts of the state (Sharamo 2014).

The west of Isiolo is the county's pastoral corridor and its most fertile part. The area is part of Oldonyiro and Burat wards, which border Samburu, Laikipia, and Meru, and is known as the livestock marketing division. Due to its fertile nature, the corridor has served as a major pastoral migration and grazing area during dry seasons, serving the Borana, Gabra, Meru, and Samburu in Isiolo County, the Sakuye and the Somalian community migrating from the east, and the Turkana migrating from the north-west. Already in colonial times, the governors of British East Africa built a slaughterhouse as a buffer zone. Müller-Mahn et al. (2021) point out that the purpose of the buffer zone was to protect the colonial settlers' cattle from livestock diseases.

140 *Evelyne Atieno Owino and Clifford Collins Omondi Okwany*

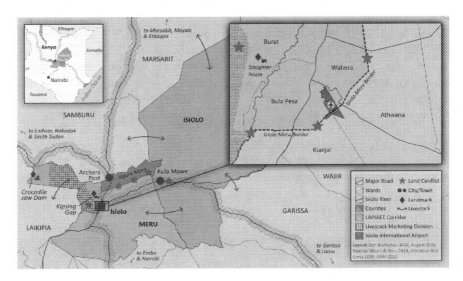

Figure 8.1 Map of Isiolo showing the contested boundary along the international airport.
Source: Map by Okwany.

In 2006, the site was chosen to erect an international abattoir, extending over 1,242 km² (Republic of Kenya 2006). The facility is planned to process about 474,000 animals annually, creating around 80 jobs and impacting at least 20,000 households (Luke 2021). Still, 18 years down the line, the abattoir is not yet completed and is estimated to be opened in March 2024. However, the fertility of the area and the linkage of the abattoir to the airport and other LAPSSET infrastructure have created contestation and competition over pasture and land among various ethnic groups.

Theory: 'frontiers' and 'state of exception'

The concept of a frontier is 'a matter of political definition of a geographical space' (Korf et al. 2013: 29). For this chapter, we use the term in its connotation of a politically contested space. State control is limited in such areas; thus, different political dynamics shape the frontier (Kopytoff 1987). Frontier spaces are characterised by a state of exception – where the government firmly applies the law when there is an emergency but generally neglects such spaces due to limited governing capacity (Agamben 2008; Korf et al. 2013; Schetter and Müller-Koné 2021). A strong military presence demonstrates territoriality – a strong control by either the government or community-based armed groups (CBAGs). However, territoriality is not just a geographical space; it is also ideological, sociological, and psychological. It is a space where the existing powers win the hearts and minds of the populace (e.g. Sack 1986). Such spaces can be strongly controlled or ungoverned frontiers, creating the ground for CBAGs or violent extremist organisations (VEOs) (Cons 2016). More specifically, territoriality can be a space with either strong government control or lack of it (Sack 1986), leading to what Okwany (2023); Okwany et al. (2023) calls semi-territoriality – the state having strong control but with limited capacity to govern, or ignoring such spaces. Thus, we adopt Agamben's (2008) thoughts of state neglect or firm exercise of force in the frontier. A strong government presence characterises such, but the state has limited security capacity or lacks interest to maintain security (Okwany 2020a, 2020b).

The limited capacity of the Kenyan government to provide security creates space for CBAGs and VEOs to operate and expand organised violence. This results in what Hansen (2013: 121–138) calls a 'win some, lose some game', where the government controls the territory in northern Kenya while militia groups exploit the state's limited capacity. Isiolo demonstrates such contestation and competition over territory and land, with new mega-infrastructure developments introducing new conflict challenges, making Isiolo a new frontier. The Kenyan state's allowance of diverse interests to create different forces, coupled with the government's dismissal of the rule of law, has enabled the exploitation and accumulation of more resources.

The government's security efforts are deficient, despite deploying the (unfit) Kenya Police Reservists (KPRs), poorly vetted militiamen employed to address the security situation in Isiolo. These KPRs were not properly renumerated and coordinated; some were involved in cattle rustling and illegal businesses. The lack of proper coordination led to the disbarment of the KPRs and the introduction of a new coordinated reservists under the National Police Service Act 2011, and the name was changed from KPRs to National Police Reservists (NPRs). Sharamo (2014) estimated that there were 4,000 NPRs in Isiolo. However, these numbers could be contested due to the untraceable recruitment and accounting processes for such NPRs and the government lacking control over their use of ammunition and movements. Consequently, the NPRs contribute to insecurity as they target and terrorise rival ethnic groups (Okwany 2023; 15.10.2022 *the Star*). Politicians also use these NPRs against their political rivals, and businesspersons and private companies hire them, while some of the NPRs join criminal networks, exacerbating insecurity (Okwany et al. 2023).

The failure of these KPRs/NPRs prompted the privatisation of security in Isiolo, which led to the establishment of the Northern Rangeland Trust (NRT), a non-governmental organisation collaborating with national and county governments in protecting conservancies. The NGO is based in pastoral land in Northern Kenya, particularly in Isiolo, and the neighbouring counties, such as Samburu and Marsabit, and it is also in the coastal Kenya, Lamu County. The NGO hires and coordinates with the National Police Service and Kenya Wildlife Service to train NPRs to protect the private conservancies in Isiolo, leading to what Lunstrum (2014: 817) calls 'green militarization'. Mkutu (2020) points out that such private security personnel (NPRs, given the name *rangers*) funded by private donors and enjoying Kenyan government support poses a risk to the county, for, if the funding ceases, the rangers would use their arms to seek other opportunities. In addition, NRT conservation aim is commercial, while pastoral communities' goal is to maintain their heritage and livelihood, this risk drawing the NGO into inter-communal conflict and local politics over the balancing and managing of pastoral grazing and conserving wildlife. Therefore, the militarisation of conservation areas and mega-developments in Isiolo shrink communal and accessible pastoral land, leading to increased competition over it.

Pastoral militarism characterises northern Kenya, where the government employs hard-power strategies, deploying military and police units, to prevent cattle rustling and banditry. In explaining the contestation and competition over land in Isiolo, we emphasise hard power as territoriality (state controlling the frontier through militarisation) and organised violence. Such strategies, however, fail due to the state's limited security resources, resulting in a form of semi-territoriality that exposes the northern frontier to conflicts. This failure can be relegated to the state's limitations to provide security and authority in the north. Armed with sophisticated weapons, cattle rustlers, and bandits, exploit this vulnerability as a method to accumulate resources (Osamba 2000). Furthermore, cattle raiding attracts criminals, with some of the county commissioners and police leadership involved in the commercialisation of such raiding (Mkutu et al. n.d.). This exacerbates violence, and the presence of transport infrastructure facilitates easy access to weapons, as road networks provide entry to porous borders in neighbouring countries such as Ethiopia, Somalia, Uganda, and South Sudan.

The land tenure system in Kenya

In tracing why things go to hell in a handbasket, we analyse land as a key variable that explains conflicts among pastoralists while sustaining their livelihoods. Historical state formation contributed to contestations and competition due to boundary-making and legal frameworks over land. The legal system governing the distribution of land in Africa, most of which is still under customary tenure, traces back to colonial administration, which introduced statutory tenure based on European property norms. Many African societies maintain their right to property without legal treatment or security (Alden Wily 2018). The rush for infrastructural development and individual property rights accelerates land grabbing and corruption (White et al. 2013). Isiolo, a hub of mega-infrastructure projects, exemplifies such dynamics.

Globally, community land is held by approximately 2.5 billion people, covering approximately 60,000,000 km² (Veit and Reytar 2017), with 20,000,000 km² located in Africa (Alden Wily 2011). This means that most African land is unregistered (Wily and Wily 2015) and under community administration. Africa's vast community land primarily includes swamplands, rangelands, or forests and are mostly neglected or considered unused (Alden Wily 2011). Only 12% of the African land is cultivated permanently (Alden Wily 2018). In Isiolo County, spanning 25,349 km² (KNBS 2019), most of the land remains unregistered and is still under communal land tenure.

Statutory tenure over Kenyan land started in colonial times with the Trust Land Act of 1938 and the African Land Development Organization of 1945. In addition, developments can be traced back to the Lyttleton Constitution of 1954 and the Lennox–Boyd Constitution of 1958.[6] Kenya's law systems predominantly derive from English common law principles, demonstrating that Acts of Parliament fall under the constitution as the supreme law. These Acts are subject to the national legislative body, which reflects the constitutional framework on land tenure.[7] Kenya

Aspirations and contestations at Isiolo International Airport   143

has 47 counties and a national government, where counties serve as trustees over communal land, most of which is to be yet registered, with only a few undergoing the registration process. Counties can enact laws to protect communal land, but the neglect of northern Kenya has hindered the implementation of protractive land laws for the population of Isiolo. The hindrance has already led to boundary conflict between county governments, and the development of infrastructure and projects along the borderland of Isiolo and Meru counties affirms such facts.

The enactment of the National Land Commission Act 2012, within the constitutional framework, established the Land Commission as an independent authority responsible for solving land disputes and administering community and trust lands. Over the subsequent years, various Land Acts were introduced (see, Table 8.1),[8]

Table 8.1 The development and subsequent land Acts in Kenya

Year	Name	Main aim of the Act
1968	The Land (Group Representation) Act	This Act gave rights and ownership of land to groups (community leadership), particularly to communities that were not hostile to the post-colonial government. The Act was synchronised into Community Land Act 2016.
1968	Trust Land Act	This Act gave rights and ownership of land to the national government. Communities that were hostile to the post-colonial governments, such as Isiolo and north-eastern counties, were under Trust Land Act. The Act was repealed by the new Acts in 2012 and 2016.
2009	National Land Policy	The document aimed to secure land rights, promote land reforms, and improve livelihoods through transparent and accountable land laws. The document was synchronised to Chapter 5 of the Kenyan Constitution.
2012	Land Registration Act	The act gave effect to the objective and principles of devolved government in land registration. It repealed the Trust Land Act of 1968.
2012	National Land Commission Act	The Act created an authority, the National Land Commission (NLC), to solve land disputes.
2012	The Land Act	This act gave an effect to Article 68 of the Constitution; it consolidated, rationalised, and revised land laws in Kenya, and it was also synchronised to Community Land Act 2016.
2016	The Community Land Act (CLA) of 2016	The law gave effect to Article 63 of the Constitution; it stipulates community land rights. It repealed the Trust Lands Act (Chapter 288) and Land (Group Representatives) Act (Chapter 287). It gave county government rights as trustees of unregistered community land. It is in parliament for amendments since 2022, due to contestation over the registration of community land. Some communities such as Isiolo believe that it will create boundaries and kill pastoral livelihoods.

collectively forming the statutory framework of Kenyan land tenure and strengthening rights for both individual and group/community ownership. Despite these legislative advancements, most of Isiolo's land remained unregistered as of 2023, despite the planning and construction of LAPSSET mega-infrastructure since 2012.

Even though the Community Land Act 2016 stipulates collective titleship, groups or communities intending to register land must also adhere to other laws, such as the Forest Act 2005 and the Climate Change Act 2016. Both acts underscore the right to protect public lands, such as forests, mountains, lakes, and other water catchments, against environmental degradation. However, in Isiolo, climate change has led to competition over land, water sources, and pastures. The politicisation of communal relations, the emergence of wildlife conservancies, and increasing changes in traditional kinship structures, weakening the traditional governance system of elders, drive conflict and violence in Isiolo (Sharamo 2014). Since colonial times, the influx of other tribes to Isiolo has also fuelled conflict over land. Mkutu et al. (2021) report that the British brought the Turkana community to Isiolo as workers, Somalians later migrated to Isiolo as traders, and the Meru from Kinna and Garbatulla took advantage of the Borana displacement, subsequently migrating to Isiolo.

The contestation over land issues in northern Kenya, particularly in Isiolo, poses significant challenges. The county government of Isiolo lacks the capacity to address land irregularities, to re-allocate land, or to protect property rights. This stirs further conflicts. The Isiolo International Airport has exacerbated these conflicts. Sharamo (2014) points out that 2,900 people were displaced in this area, and 165 were killed from 2009 to 2013 following the Ngaremara ward boundary dispute between Borana and Turkana. Mkutu and Boru (2019) argue that the airport complicated ethnic relations between Isiolo and Meru because it fuelled land conflicts.

The rivalry over land escalated into conflicts involving both investors and communities, as well as the state and communities, evolving into a protracted ethnopolitical conflict. Nomadic pastoralists, such as Borana and Somalis, view Meru as the ethnic community favoured by the government for its claims over land in the project areas. This created speculation about favouritism, leading to ethnopolitical conflicts (Greiner et al. 2022). The bias is further evident from the outset of the project in 2012. Some members of LAPSSET's high-level authority belonged to the Meru ethnic community, while the pastoral communities lacked political representation to communicate their interests to the government.

Mkutu and Boru (2019) underline that both the Meru and Isiolo communities lost their land to the international airport, constructed on trust land – communal land under customary tenure before the enactment of the 2012 Land Act and the CLA of 2016, designed to protect the pastoral communities from land grabbing. Müller-Mahn et al. (2021) affirmed that there was a delay and slow implementation of these laws, coupled with the rapid changes brought by the mega-infrastructure projects, leading to contestation and competition over the land.

While the Meru community in the Nyambene area bordering Tigania East and West Subcounty was compensated, the pastoralists on the Isiolo side in Wabera ward were not compensated due to the lack of land titles in Isiolo County (Owino

2019). Consequently, the competing aspirations and contestations and assumed favouritism experienced by pastoral communities during the airport expansion triggered historical grievances of social injustices towards the northern frontier communities following the colonial legacy and past government policies. Exploiting this situation, the CBAGs and VEOs capitalised on recruitment efforts to address grievances, contributing to propaganda and radicalisation of Kenyan youth (Hansen et al. 2019; Okwany 2016, 2023). Since 2013, about 200 youths from Isiolo have been recruited to join the Somalia-based VEO (17.05.2019 *Reuters*). Al-Shabaab[9] has spread claims about land grabbing and grievances, and the 2019 Riverside Drive attack at the Dusit2 complex in Nairobi, which killed about 21 people, was planned by a former Isiolo resident, Ali Salim Gichunge, who was a son of a military officer of the Kenya Defence Forces (17.05.2019 *Reuters*).

LAPSSET and its connection to the airport

LAPSSET includes roads, railways, ports (32 berths), pipelines, fibre-optic connections, and mega dams, connecting to international airports in Isiolo, Lake Turkana, and Lamu County. The transport corridor is anticipated to connect Kenya's international waters of Lamu, passing the arid areas from Garsen in Tana River County to Isiolo. Upon completion, Isiolo will be situated at the LAPSSET junction, connecting Ethiopia through Marsabit-Moyale to Addis Ababa. It aims to link Isiolo International Airport to these areas (see Figure 8.1), extending to Samburu to the Turkana Nakodok border via Lodwar and ending in Juba, South Sudan (Okwany 2020c). A shift in development focus towards northern Kenya emerged in the 1990s, when Isiolo's land and town ownership started evolving along the pastoral corridor. In 2012, the launch of the LAPPSET triggered the international transportation corridor, with the anticipation of improved regional connectivity transforming Isiolo from an unworthy or unproductive region to a more developed municipality. The expectation of a well-connected and spatially integrated region created speculation of land and led to land grabbing (see later text; also Elliott 2016; Owino 2019). In February 2012, a land office was established in Isiolo for the first time since Kenya's independence to facilitate the land use processes in relation to the anticipated LAPSSET. One month later, a LAPSSET proposal meeting was arranged between the late president Mwai Kibaki of Kenya, former Ethiopian prime minister Meles Zenawi, and South Sudan's president, Salva Kiir Mayardit (Browne 2015).

The LAPSSET Corridor Development Authority (LCDA) report affirmed that the mega-infrastructure is expected to boost 2–3% of Kenya's gross domestic product (GDP), supporting the government's ambition of an 8–10% economic boost when completed (LCDA 2016). By 2010, LAPSSET speculations had already attracted international contractors and donors, such as China, through its Exim Bank's rural electrification initiative (Okwany 2020c). In this initial stage, LAPSSET was promising to attract a US$12 million grant to the corridor, while the Chinese engineering firm CAMC also committed to upgrading the power plants in Lamu from a single-circuit phase of 33 kV to 220 kV (Bremner 2013). However,

LAPSSET's envisioned multilateral partnership was facing challenges. Initially attracting countries such as India, Brazil, Qatar, South Korea, and China, as well as the European Union, these partners pulled out due to the drop in global oil prices in 2013–2014 and uncertainties surrounding oil in South Sudan and Turkana in Kenya. Consequently, the Chinese government became the sole international partner (Okwany 2020c). In addition, LAPSSET faces competition from other regional mega-infrastructure, such as the Djibouti–Ethiopia road, port, and railway project and the pipeline from coastal Tanga in Tanzania to Hoima in Uganda (Okwany 2020c).

After the completion of the airport, economic aspirations and political ambitions fanned by the LAPSSET corridor, as outlined in the Kenya Vision 2030 reports (2018, 2020), faced controversies on the ground. The airport is operating below its intended capacity of handling 125,000 passengers annually, has encroached upon the pastoral grazing corridor, and is showing little economic value to pastoral livelihoods. The airport was expected to boost the pastoral meat market economy through streamlined and faster transportation (04.07.2021 *Daily Nation*). However, this intended use has stalled because of the delay in the abattoir's development.

Furthermore, the UK government's ban of khat imports in 2014 was another factor that led to the economic failure of the airport, and the flights from Isiolo to Nairobi are expensive in comparison to those from Nairobi to Lodwar or Nairobi to Kisumu, which cover larger distances than the flight from Nairobi to Isiolo (Owino 2019). The airport's slow operation is also linked to the politics of anticipation, which overshadow the LAPSSET corridor. The LCDA affirms that the airport land has been expanded from 1.5 km² to occupying 2.59 km² of contested land between fertile Meru County and the arid part of Isiolo (LCDA 2017).

Findings on the local consequences of airport expansion

Airport-induced displacements, irregular compensations, and challenging land re-allocation

The expansion of the international airport began in 2004 with the construction of a 1,400 m runway. The plan was to expand it to 3,000 m. The Kenya Airport Authority (KAA) and a village council of elders from Meru and Isiolo passed an agreement to expand the airport and to assess impacts on the local population prior to compensation.[10] Both councils and the KAA agreed to resettle people affected by airport expansion.[11] Resettlement began in July 2008 in the Mwangaza area (Wabera ward) close to the airport. The resettlement resulted in many legal landowners losing their land due to land fraud in the Kiwanjani and Mwangaza areas, such as double allocation of allotment letters (titles in trust land), and those with power, influence, and money manipulated the system to get land, while community members with no influence lost their land.[12] Issues such as the delay and slow implementation of land policies and legislation led to the flouting of the agreement.[13] For example, a National Land Commission officer in Nairobi indicated that resettlement was stopped because the new Kenyan Constitution from 2010 overruled the 2004 agreement between the KAA and the councils of elders. The officer told the authors:

A new county government is in place with new laws . . . , and such land complaints are difficult to solve without land registration. Furthermore, the airport was already approved and completed, and flights are in operation.
(National Land Commission officer, 21 February 2019)

The agreement with the nomadic pastoralists was dismissed on the premise of a new Isiolo County administration that had introduced new land laws. The land in the Nyambene area, Meru County, is privatised land and not communally owned, like in Isiolo County. Community Land Act (CLA) 2016, for example, deals with land laws that are not older than 2010. Therefore, agreements with the airport authority, such as that of resettlement, are deemed not legally binding.[14] Consequently, airport expansion displaced people, many of whom were forcefully evicted. While the KAA claimed to have compensated the affected communities, some dispute this claim. In addition, the airport expansion led to the unequal destruction of private properties and loss of livelihoods due to territoriality – state control of the airport land area. Organised violence – state monopolising the use of force or ethnic communities – also legitimised the use of force due to eviction from their land. For example, a public school was demolished. Pupils had to attend schools much farther away, disrupting their learning activities. Over 1,300 evicted people lost their land and property.[15]

There are court orders over land adjudication in contested boundaries, and a parliamentary commission is expected to look into these disputes. For example, Mwangaza village has land cases pending in court to know who should be compensated.[16] The most contested land areas are Mwangaza, Kambi Juu, Kambi Gabra, Chechelesi, and Kiwanjani. These regions are characterised by double registration of land and tribal conflict caused by infrastructural developments.[17] In addition, our investigation of the court cases at the Isiolo Court demonstrates that land claims increased from 10 cases in 2016 to 30 cases in 2017 and 92 cases in 2018.[18] The sudden increase was due to the expansion of the airport since 2011, and the anticipation of the proposed mega road since 2012, and the rush to create the land office in the same year. According to the Isiolo Town Council's report of 2005, the airport's expansion displaced 1,337 people in Kiwanja-Ndege and Mwangaza.[19] However, data from the National Land Commission, the KAA, and the Isiolo County Commissioner's Offices list the number of displaced persons at 64. This official list, therefore, is contested. Our evidence indicates a much larger number of squatters in Mwangaza and Kiwanja-Ndege, estimating, indeed, more than 1,300 displaced people.[20]

Consequently, communities in Isiolo claim that the Meru communities do not respect such processes and that the Meru County government is adjudicating the contested land in favour of its population.[21] As mentioned earlier, in Kenya, rural landownership is primarily regulated by customary law, and these communal lands were registered under the trust land. However, the contention lies in the trustee issues, which have been associated with corruption cases. Before the promulgation of the Kenyan Constitution, trustees (custodians) of unregistered community land were town councils who sold land to private entities and accumulated private land for their own benefits.[22] However, since the Community Land Act of 2016, there has been progress in some counties in community land registration, but this process faces challenges due to customary practices.

In Isiolo County, the titleship for community land started in 2018, after the parliamentary legislation on land regulation in 2016 and executive order No. 1 of 2018 mandating the Ministry of Lands and Physical Planning to adjudicate and register land under community membership.[23] The year 2022 demonstrated the highest land registration of Isiolo land under the national titling program.[24] The Constitution of Kenya stipulates that there should be no discrimination over land registration. Article 63 of the Constitution and Community Land Act 2016 stipulate community land rights.[25] Communities in Isiolo have experienced conflicting debates over land despite a clear legal framework of such land laws; there is limited engagement and consultation between the communities in Isiolo and the national/county government over community land administration.[26]

People in Kiwanjani and Mwangaza were not compensated because they lacked land title deeds, as their land became subject to the Trust Land Act under the custodianship of the former Isiolo Town Council, now Isiolo County. These lands were not registered before the Community Land Act of 2016, and with the law in place, registration is still due at the time of writing. The land victims from Tigania East and West in Meru had title deeds; thus, their compensation was easy.[27] The pastoral communities have grievances over compensation and reported the Meru communities demarcating and claiming pastoral land to the chiefs, security teams, and county commissioner, yet little action was taken to ease the tension.[28] When we wrote this chapter, Isiolo County had no land registrar, and most of the land was not registered.[29] However, under the national titling program, the government, through its National Treasury's cabinet secretary, Ukur Yatani, issued 6,000 title deeds to residents in Odonyiro, the location of the abattoir, and Ngaremara, the proposed area for the LAPSSET junction.[30]

In July 2019, the parliamentary senate committee claimed that the international airport was idle. Yet the government spent 2.7 billion Kenya shillings on its upgrading, and the National Land Commission failed to compensate individuals whose lands were taken by the government to upgrade the airport.[31] Residents near the airport claim such irregularities and a lack of the Kenyan government's commitment to compensate those who lost their land.[32] Due to corruption allegation, land conflict, and double allocation of land, the national government created a balloting exercise to solve land grievances in the Wabera ward. However, the exercise faced numerous irregularities, including a corrupt deal because of unregistered pastoral land. Half of Wabera ward's land lacked titles, as it is communally owned, rendering it vulnerable to appropriation by state and county politicians.[33] Such irregularities intensify ethnopolitical conflicts, because unkept government promises, lack of or skewed compensation attract political rhetoric, particularly during election campaigns.[34]

More so, locals claim that government officials, including county commissioners and senior police officers in the region, have been allocated plots, anticipating the LAPSSET and the operation of the airport.[35] As such, only the elites can acquire substantial land portions, leading to practices such as land banking, manipulation, and speculation, while for local communities, obtaining titles and claiming their land have become a challenge.[36] Land grabbing became a prominent political issue for opposition parties seeking voters in the 2017 election year. However, the political handshake between the president and the opposition leader slowed the land-grabbing debate.[37]

Local political struggles

As mentioned earlier, ongoing contestation and competition over land persist around the airport ground, leading to escalating ethnopolitical conflicts and insecurity in the LAPSSET region. Land is viewed as state territory, and this view can sometimes conflict with traditional community and individual landownership, particularly when the state's economic interests clash with communal or individual land use. Even though the Community Land Act of 2016 registered about 32 plots of community land, the registration is not entirely new, because these community lands were registered under the Trust Lands Act and the Land Group Representation Act.[38]

Mistrust between the state and the community over land use and development prevails, with the state being viewed as exerting its power over the community. The contestation over the Kenya Defence Forces (KDF) land exemplifies community scepticism when the state acquires communal land for development projects.[39] In 1981, the military and the Isiolo communities' leadership had an agreement on a military training space, allocating 106.53 km² of training land in Isiolo to the military. However, the military has been territorialising, that is, controlling, the space and applying organised violence, legitimising the use of force to expand and fence more land. This has caused conflict between nomadic pastoralists and KDF over grazing rights. Furthermore, the disposal of hazardous military waste is a risk to the community's well-being.

The airport was established on disputed land between Wabera ward in Isiolo and the Nyambene area bordering Tigania East and West constituencies in Meru County. This led to border conflicts between the Meru and Isiolo,[40] as the airport occupies 20% of Wabera land, with 80% of its runway situated in the Nyambene area, claimed by Meru County.[41] Wabera ward, under Isiolo Town municipality, has a history of land conflicts between Borana and Samburu, Borana and Meru, Somali and Samburu, Turkana and Meru, as well as Turkana and Samburu communities. The area where the airport stretches is not only a disputed land between Meru farmers and Isiolo pastoralist but is also characterised by the county boundary dispute and is prone to cattle rustling.[42]

The extension of the airport resulted in Isiolo nomadic pastoralist communities becoming landless squatters on their own land. This exacerbated their situation, as they perceived that the agriculturalist and Christian Meru had been favoured.[43]

Thus, the future of the airport has triggered community resistance, leading to various consequences, including boundary issues.[44] Such perceived favouritism led to the marginalisation of some communities, making Isiolo a prime target for VEOs, such as al-Shabaab, to find recruits. These VEOs, but also community-based armed groups (CBAGs), exploit these land and religious grievances to radicalise the already-vulnerable youth.[45] Such radicalisation and recruitment are common and increasing in the Wabera ward.[46] The expanded road network poses an increased risk of small and light weapons trafficking and the likelihood of future insurgency threats.[47]

Our research in 2018 and 2019 demonstrates that al-Shabaab has recruited more than 200 youths from Isiolo. The VEO exploits the land-related grievances experienced by the pastoralists in Isiolo, the majority of whom are Muslim. It also takes advantage of Isiolo's semi-territoriality, where the state exerts strong control but has limited capacity or overlooks some spaces. This situation becomes advantageous to al-Shabaab, leveraging land grievances and religious propaganda to win and recruit youth in Isiolo County.[48]

Similar to the conflicts in the Wabera ward, the Ngaremara ward also faces contestation and competition over LAPSSET. This ward is home to Turkana and Borana nomadic pastoralists, while the Gare clan from Somalia migrates to the area for business. Meanwhile, the Meru ethnic group claims that the ward is Meru communal land and falls under Meru County's jurisdiction. These dynamics lead to inter-ethnic conflicts. The Turkana pastoralists reside in the Nakuprat location, and the Borana inhabits the Gotu area, which is part of the Ngaremara ward. The Gare clan are a business community, purchasing land in Gotu, hoping to profit from anticipated future development opportunities, thus causing land conflict with the Borana. Ngaremara ward also hosts a community conservancy, Nakuprat-Gotu, where both Turkana and Borana ethnic communities share the conservancy.

The Northern Rangeland Trust (NRT) protects the conservancy with the help of about 16 rangers, 8 from Turkana and 9 from Borana. The community conservancy was created as a buffer zone to prevent the Turkana and Borana from raiding each other.[49] Thus, the conservancy stands as one of the NRT's successful models of community conservation and policing, engaging both Turkana and Borana in peace initiatives and sharing resources with the Nakuprat-Gotu community conservancy.[50] However, there is still contestation over land in this conservancy: the Meru community claims ownership, evident through painted stones and trees demarcating conservancy areas as their territory.[51] Land disputes revolve around cattle rustling, business competition, and land grabbing in anticipation of the upcoming LAPSSET project. Both the Borana and Meru political elites are also additional actors in the contestation, claiming part of the conservancy.

With NRT rangers being well trained and equipped, there is a concern that they might use their sophisticated arms for economic purposes if faced with a cessation of donor funding. National Police Reservists (NPRs) are also present in Ngaremara ward. There are about 17, and they help in community policing. Distinctively, NPRs focus on safeguarding the community against livestock theft and poaching, while NRT rangers are tasked with protecting community conservancies.[52]

However, there are also risks of these groups joining criminal groups or supporting their ethnic groups during ethnic clashes, because of poor remuneration and a lack of government accountability.[53]

Over the years, Kachuru and Kula Mawe/Kulamawe areas have also experienced boundary conflicts.[54] Kachuru and Kulamawe are located approximately 60 km and 73 km from Isiolo airport, respectively. A majority of Borana, Somali, and Meru ethnic communities occupy these areas.[55] The LAPPSET road is estimated to pass both Kachuru and Kulamawe, and there is limited security in these areas, leading to conflict among the Meru and Borana, and the VEO could take advantage of such insecurity. Administratively, Kachuru and Kulamawe are in Meru County but claimed by the Borana; more so, since the implementation of devolved government in Kenya, boundary conflict between Meru and Isiolo counties has escalated in these areas. The proposed resort city, another component of the LAPSSET project, will be located in the Kipsing gap region. Previously, the Kenyan government had proposed the Oldonyiro area for the city but later dismissed it. The Meru are pursuing and claiming the Kulamawe and Kachuru area (see Figure 8.1) because of the proposed resort city. This is leading to yet further land contestation and competition.

Discussion and conclusion

Since the Kibaki administration's second term in 2008, the promulgation of the Constitution of Kenya in 2010, and the introduction of devolution in 2013, development projects in Isiolo have blossomed. Interviews with county government officials confirmed that Isiolo and the northern frontier had indeed been the neglected 'Kenya B' – an unproductive or unworthy region. But this perception is changing. The funds allocated to the Isiolo government in six years (2013–2019) through devolution is more than what the region received in nearly 50 years (1963–2012).[56] But despite the increased focus on the north due to projects like the airport and LAPSSET, some developments, such as the international airport, have not fully realised their promised impact.

Some Isiolo County officials and politicians claim that local communities were not involved in public participation during project planning,[57] and communities interviewed in this chapter confirmed the same. In addition, they argue that most communities in Isiolo and nearby areas are poor and cannot afford to transport their horticultural products and meat from Isiolo to Nairobi. The communities' needs and economic progress were not considered in the planning. This shows that there is a disconnect between community interests and the national government's concept of development.[58] Similar to the British colonial government, the Kenyan government officials from Nairobi are subjecting rural communities to development projects that are economically redundant, thus leading to conflicts over differing visions of development.

The future of Isiolo is uncertain despite the Kenyan government's exaggerated development plans. The county is poised to experience exacerbated ethno-political conflict and violence, including contestation between government development projects that shrink pastoral grazing land, leading to armed conflict over pasture and water[59] due to unmet community needs. The international airport's economic projection has stalled, with fewer flights than expected. Yet there are still speculations and anticipation over land because of the proposed LAPSSET (Enns 2019). While these projects aim to shape Isiolo with increased tourism and easier transportation of khat to Nairobi and other international markets (Mkutu and Boru 2019; Owino 2019), such development also means an uncertain future marked by land disputes, turning Isiolo into a new frontier of conflicts. The airport, which is expected to serve as the transportation hub of meat from the proposed abattoir and khat from Meru, is facing economic challenges because the transportation of flowers, khat, and meat is cheaper via road than by air.[60] In addition, flights from Isiolo to Nairobi are expensive in comparison to other distance, such as Nairobi to Lodwar or Nairobi to Kisumu (Owino 2019).

The Kenyan government has faced delays in the progress of its large-scale projects in Isiolo. Yet the hyped development creates speculations and anticipation, with people still buying land, hoping to reap economic benefits when the projects are completed, and Isiolo undergoes development. A newspaper report (19.07.2020 *Daily Nation*) affirms that even after receiving 800 million Kenyan shillings from the World Bank, the Isiolo abattoir faced several delays, pushing its completion deadline from December 2016 to February 2021. And in December 2023, the abattoir was yet to be operational. The slaughterhouse is envisioned to process about 474,000 animals yearly to boost the pastoral economy, aiming at markets such as the Middle East. It was also to breathe new economic life into Isiolo International Airport. However, the delays signify future uncertainties for the abattoir and the international airport. Such uncertainties are also evident in the proposed LAPSSET project, which has faced challenges of international partnerships (Okwany 2020a). Yet the Kenyan government had high ambitions over LAPSSET, aiming at an 8–10% increase in the country's GDP (LCDA 2016), and such hopes are still maintained. Essentially, it should be noted that even with these ambitions, the future of the nomadic pastoral communities is uncertain as government initiatives on compensation appear to favour the Meru ethnic community and marginalise the pastoralists.[61] As such, their grievances over government marginalisation and the Meru background of some top officials in the LAPSSET authority continue to exacerbate the relationship between the state and society (Greiner et al. 2022; Owino 2019). The number of court cases over land in Isiolo has risen recently, especially within the Wabera ward, where the Meru community that continues to claim land in Ngaremara, Kulamawe, and Kachuru areas keeps ignoring complaints of the pastoral communities.

Just as the colonial British ignored the Northern Frontier District, the Kenyan government ignored northern Kenya, considering it unworthy of attention, until the 1990s (Elliott 2016). The Land Act of 2012 and the Community Land Act of 2016 changed this and protect communities and individuals from land grabbing. However, implementation is slow, while development projects are advancing rapidly (Müller-Mahn et al. 2021). Despite the constitutional framework of land rights, the Kenyan government exhibits a disregard for the rule of law, with land laws not being respected and compensation favouring some elites and the Meru ethnic community.[62] Such dismissal of the rule of law is what Agamben (2008) and Korf et al. (2013) call a state of exception, where the state has the power to apply or suspend the law, has limited capacity, or ignores the law due to some interests, leading to conflict and violence in the frontier (also see Kopytoff 1987). Despite the state of exception in Isiolo, where the government allows the military to use land and subject the Isiolo communities to conflict and violence but also applies the law in case of conflict among the ethnic tribes in Isiolo, there has been increased insecurity due to what we describe as territoriality, the use of hard power/control, and organised violence, and thus military legitimizing the use of force. Also, there has been a rush to have an administrative land office put in place. Such a rush has led to land disputes in court, some people having their names on the same land but different titles, while government officials acquire land in strategic positions, speculating and anticipating development that is yet to be realised.[63]

Essentially, Isiolo has a history of land disputes, cattle rustling, and pastoral and ethnic conflicts. Land disputes arising from development projects such as the international airport and LAPSSET lead to future contestation, conflict, and dynamics of violence. The future of Isiolo's security is uncertain, and there is the concern that new actors such as VEO and CBAGs that take advantage of the land grievances, recruiting and radicalizing young people, will contribute to future insecurity and the risk of insurgency (Okwany 2016, 2023).[64] Our fieldwork research in 2018 and 2019 demonstrated that al-Shabaab recruited more than 200 youth from Isiolo County using land grabbing and grievances in their recruitment propaganda.[65] Even though the road network in Isiolo is advantageous to the county because of movements of goods and services, Mkutu and Boru (2019) point out that the porous borders in the north towards Moyale and the long border with Somalia risk the proliferation of small arms and light weapons.

The contestation and competing aspirations over land in Isiolo are not new; the county is characterised by ethnic and boundary conflict, land disputes, and cattle rustling. However, the hyped mega-infrastructure projects escalate existing conflicts due to further fragmentation of pastoral rangelands. Isiolo International Airport was completed in 2017; however, some renovations and its development into an international standard airport are yet to take shape. In addition, the airport faces economic challenges as aircraft companies withdrew from flying to Isiolo due to limited traffic, low demand, and expensive travel tickets. At the time of writing, the airport project is not viable to the local economy, and it is yet to achieve its primary goal of transporting horticultural products, such as flowers and khat, including livestock products, from the proposed modern abattoir. The international airport is facing land disputes; pastoralists who were evicted from their land in Wabera ward, some of whom have become squatters, are yet to receive compensation. In contrast, the Meru community members evicted from their land in the Nyambene area have been compensated. This state of affairs leads to ethnopolitical contestations.

Based on our findings, the completed airport and the abattoir, which were expected to upgrade pastoral livelihoods and boost the country's economy; the proposed resort city; and other LAPSSET projects, such as roads and the proposed mega-dam, create speculations and politics of anticipation. The Meru community is claiming the land in Ngaremara, Kulamawe, and Kachuru areas, which have been identified to benefit from the corridor. The elites are also securing land in these areas, with the anticipation of commercialisation of land, while the local nomadic pastoralists have no titles due to their nature of migration and search for pasture. Furthermore, the airport, abattoir, resort city, and road are supposed to boost Isiolo's economy, yet the community is not involved in such projects. With the state applying a top-down approach towards development in the northern frontier, these affected communities in Isiolo are resisting the projects, leading to different dynamics of conflict and violence. The history of militarisation, violence, and marginalisation in Isiolo risks passing the county's future to CBAGs and VEOs. Therefore, Isiolo has emerged as the new frontier of contestations and competing aspirations.

Acknowledgments

The research was part of the project Future Rural Africa funded by the Deutsche Forschungsgemeinschaft (DFG, grant TRR 228/1 2018–2021) at the University of Bonn, which also funded open access. The research also benefited from the completed EU-Horizon 2020 research and innovation project ICT4COP, a Community-Based Policing and Post-Conflict Police Reforms project hosted by Norwegian University of Life Science.[66] Interviews originating from ICT4COP are indicated. The authors thank Hanna Geschewski and Irit Ittner for their editorial work.

Notes

1. Aircraft company management team member, 04.02.2019, interview.
2. Businesspersons in Isiolo and Meru, 27.09.2028/16.02.2019, interviews.
3. Kenyan government rhetoric through the spokesperson, Rtd Col. Cyrus Oguna, to the Kenyan mainstream media in 2021.
4. The conference was organized by the Kenyan Ministry of Land and Physical Planning, French Institute for Research in Africa, and Food Agriculture Organization of the United Nations, 13-14.06.2022.
5. See, for example, session papers No. 10 of 1965, and No. 1 of 1986 Republic of Kenya (1965). *Sessional Paper No. 10 of 1965 on African Socialism and its Application to Planning in Kenya*. Nairobi: Government Printers. Retrieved from https://repository. kippra.or.ke/handle/123456789/2345, Republic of Kenya (1986). *Sessional Paper No. 01 of 1986 on Economic Management for Renewed Growth*. Nairobi: Republic of Kenya. Retrieved from https://repository.kippra.or.ke/handle/123456789/2679.
6. The Lyttleton Constitution 1954 and Lennox–Boyd Constitution of 1958 mainly focused on adding the African race to the Legislative Council (LegCo). The main debate for the LegCo was African property and land; thus, community representation and property rights were key (see The Round Table. (1958). A constitution for Kenya: Mr. Lennox-Boyd's visit. *Taylor and Francis* 48(190). https://doi.org/10.1080/00358535808452115. However, the post-independence government further developed such a statutory framework by enacting the Land Group Representation Act 1968, Trust Land Act 1968, Land Consolidation Act 1977, Land Adjudication Act 2010, and the 2009 land policy, which was synchronized into the Constitution of Kenya. (2010). Nairobi: Government Printers Retrieved from http://kenyalaw.org/kl/index.php?id=398.

7 See Article 63 of the Constitution of Kenya on community land.
8 See Land Registration Act 2012, the Land Act 2012, and the Community Land Act (CLA) of 2016; this legislation repealed the Trust Land Act of 1968, while the Land (group representation) Act 1968 and Land Act 2012 were synchronized into CLA in 2016.
9 A Somalia-based VEO and concentrating its attacks in the Horn of Africa and affiliated to Al-Qaeda, a Salafists jihadists group, associating to the Sunni Islamist ideology.
10 Letter written by the KAA to the Meru and Isiolo council, 17.09.2004.
11 Three representatives, Wabera Ward Administration, 24.01.2019, interview.
12 Squatters in Kiwanjani Zone G, 10.09.2018; ward administrator, 23.01.2019, interviews.
13 People displaced by the airport expansion, 25.01.2019, FDG.
14 A national land commissioner and a lawyer, Ministry of Land and Physical Planning, 14.06.2022, interviews.
15 A county government officer and a county executive commissioner of tourism, Isiolo, 27./28.08.2018, interview.
16 Chief of Mwangaza, 09.09.2018 and 14.03.2019, interviews.
17 Ward administrator affected by the airport expansion, 25.01.2019, interview.
18 Executive officer, Isiolo law court, 25.01.2029, interview.
19 Isiolo senior national government administrators, 28.10.2018 and 24/25.01.2019, interviews.
20 The numbers of displaced people were retrieved from unpublished documents during our interview with Wabera Ward Administration, 24.01.2019. Leaders of displaced people in Mwangaza also confirmed the numbers, 24.01.2019 and 14.03.2019, interviews.
21 Isiolo County official, 11.02.2019, interview.
22 A ward administrator and three members of County Assembly demonstrating land corruption in Isiolo before dispensation of devolved governance, 09. and 17.09.2018, interview. A land commissioner and two land administrators at the conference on Community Land Act 2016, 13.06.2022, conversation.
23 See executive order no. 1 available at https://lands.go.ke/about-us/.
24 A land registrar from Laikipia and a land adjudicator from Samburu, 17.08.2022, interview.
25 Laikipia land registrar and Samburu land adjudicator, 13/14.06.2022, interviews.
26 Secretary of Pastoralist Parliamentary Group and chief executive officer of the Drylands Learning and Capacity Building Initiative, 15.06.2022, interview.
27 FGDs in Chechelesi, 08.09.2018; Kiwanjani, 18.10.2018; Kambi juu, 19.10.2018.; Kambi Gabra, 12.03.2019; Mwangaza, 25.01.2019.
28 Chairman of Mwangaza, from where people were displaced, 22.01.2019, interview.
29 A land registrar from Laikipia and a land adjudicator from Samburu, 17.08.2011, interview.
30 The National Treasury cabinet secretary, Ukur Yatani, issued 6,000 land titles publicly on 11 July 2022 to Ngaremara and Odonyiro residents, increasing the number of land titles registered in Isiolo to 6,300.
31 KTN News, *New Roundup, Isiolo idle after shs.2.7billion used to upgrade it*. Available at https://youtu.be/-sg2gcqzzxc.
32 Isiolo County officer, 12.03.2019, interview.
33 Chief in Wabera ward, 15.09.2018, interview.
34 Fieldwork evidence on the electioneering period over the land conflict from April to July 2022; politicians used the land conflict as a campaign tool in Isiolo.
35 Villagers in Mwangaza, 13.09.2018, interview.
36 Habiba, Wabera ward, 19.09.2018, interview.
37 The opposition leader, Hon. Raila Amollo Odinga, claimed that powerful figures from the ruling party, Jubilee, were grabbing land in Ngaremara, speculating on the emergence of the proposed LAPSSET junction and its connection to the airport. However,

in 2018, after Raila was appointed the African Union high representative for infrastructure development in Africa, he changed his opinion on land grabbing in Isiolo, now claiming it would attract donors to the LAPSSET initiative locally and internationally, 13.02.2019, interview.
38 A national land commissioner and a lawyer at the Ministry of Land and Physical Planning, DATE, interview.
39 Information shared during the conference on Community Land Act, 13/14.06.2022.
40 Residents in Mwangaza, 17.09.2018, interview.
41 Residents in Mwangaza, 18.09.2018, interview.
42 Chief in Wabera ward, 20.09.2018, interview.
43 Security fraternity and communities in Isiolo town, 16.-21.04.2018, interviews/FGDs, ICT4COP project.
44 Elders in Ngaremara ward, 07.09.2018, interview.
45 Mr Maiyo, deputy county commissioner, 28.08.2018, interview.
46 Habiba, the Wabera ward administrator, 19.09.2018, interview.
47 Isiolo Peace Committee, 19.10.2018, FDG; assistant county commissioner Isiolo, 28.08.2018, interview.
48 Residents in Ngaremara, and Wabera ward, 02.2018; 11.2019, interviews, ICT4COP project.
49 Nakuprat-Gotu community wildlife conservancy rangers, 18.09.2018, interview.
50 Mrs Ekiru, the NRT peacebuilding coordinator (northern Kenya), 06.09.2018, interview.
51 Observation, 09.2018 and 02.2019.
52 NTR council of Elders, 06.09.2018, interview.
53 Five NPRs in Ngaremara, 02.02.2018, interview.
54 Isiolo county officer, 22.02.2019, interview.
55 Elders in Ngaremara ward, 07.09.2018, interview.
56 Isiolo county officer, 22.01.2019, interview.
57 Mrs Habiba, Wabera ward administrator, 19.09.2018; Isiolo County officer, 22.01.2019, interviews.
58 Mr Akall, board member of the Dry Drylands Learning and Capacity Building Initiative, 14.06.2022, interview.
59 Ngaremara and Wabera wards, 02.02.2018/23.01.2019, interviews demonstrating concerns over LAPSSET, compensation, and Isiolo International Airport shrinking pastoral land.
60 Businesspeople in Isiolo and Meru, 09.2018, 02.2019, interviews.
61 Mwangaza Chief, 14.03.2019, interview; Mwangaza community members and chairperson and Isiolo county officials, 15.03.2019, interviews.
62 Mwangaza community members and chairperson and Isiolo county officials, 15.03.2019, interviews.
63 Mwangaza chief, 14.03.2019; Isiolo County officials and an executive officer at the Isiolo law court, 25.01.2019, interviews.
64 Mr Maiyo, deputy county commissioner, community members, and police in Isiolo central in 2017, 2018 and 2019, interviews, ICT4COP project.
65 Residents in Ngaremara, and Wabera ward, 02.2018, 11.2019, interviews, ICT4COP project.
66 Norwegian University of Life Science. "Community-Based Policing and Post-Conflict Police Reform: A European Commission Horizon 2020 Research & Innovation Project, Ict4cop Research Project." 2020. https://cordis.europa.eu/project/id/653909/results. Accessed 07.12.2020.

References and sources

Agamben, G. (2008). 'State of exception'. In *State of exception*. Chicago: University of Chicago Press.
Alden Wily, L. (2011). *The tragedy of public lands: The fate of the commons under global commercial pressure*. Rome: Contribution to ILC Collaborative Research Project on Commercial Pressures on Land.

Alden Wily, L. (2018). 'The community land act in Kenya opportunities and challenges for communities'. *Land* 7(1), 12.
Bremner, L. (2013). 'Towards a minor global architecture at Lamu, Kenya'. *Social Dynamics* 39(3), 397–413.
Browne, A. J. (2015). *LAPSSET; The history and politics of an eastern African megaproject*. Nairobi: T. Rift Valley Institute.
Cons, J. (2016). *Sensitive space: Fragmented territory at the India-Bangladesh border*. Seattle: University of Washington Press.
Constitution of Kenya (2010). Nairobi: Government Printers. http://kenyalaw.org/kl/index.php?id=398
Elliott, H. (2016). 'Planning, property and plots at the gateway to Kenya's 'new frontier''. *Journal of Eastern African Studies* 10(3), 511–529.
Enns, C. (2019). 'Infrastructure projects and rural politics in northern Kenya: The use of divergent expertise to negotiate the terms of land deals for transport infrastructure'. *The Journal of Peasant Studies* 46(2), 358–376. https://doi.org/10.1080/03066150.2017.1377185
Greiner, C., S. Van Wolputte, and M. Bollig (2022). *African futures*. Leiden: Brill.
Hansen, S. J. (2013). *Al-Shabaab in Somalia: The history and ideology of a militant Islamist group*. Oxford: Oxford University Press.
Hansen, S. J., S. Lid, and C. C. O. Okwany (2019). *Countering violent extremism in Somalia and Kenya: Actors and approaches*. Oslo: OsloMet.
Kenya National Bureau of Statistics (KNBS). (2019). *2019 Kenya population and housing census*. Nairobi: KNBS.
Kenya Vision 2030 Report (2018). *Kenya vision 2030 flagship projects progress report 2008–2017: Towards a globally competitive and prosperous nation*. Nairobi: Kenya Vision 2030. https://vision2030.go.ke/publication/kenya-vision-2030-flagship-programmes-and-projects-progress-reports/
Kenya Vision 2030 Report (2020). *Kenya vision 2030 flagship projects progress report 2018–2020: Towards a globally competitive and prosperous nation*. Nairobi: Kenya Vision 2030 Secretariat. https://vision2030.go.ke/publication/kenya-vision-2030-flagship-programmes-and-projects-progress-reports/
Kopytoff, I. (1987). *The African frontier: The reproduction of traditional African societies*. Bloomington/ Indianapolis: Indiana University Press.
Korf, B., T. Hagmann, and M. Doevenspeck (2013). 'Geographies of violence and sovereignty: The African frontier revisited'. *Violence on the Margins*, Springer, 29–54. https://doi.org/10.1057/9781137333995_2
LAPSSET Corridor Development Authority (LCDA). (2016). *Brief on LAPSSET corridor project*. Nairobi: LCDA.
LCDA (2017). *Consultancy services for the strategic environmental assessment for the LAPSSET infrastructure corridor*. Nairobi: LCDA.
Luke, B. (2021). 'Opening of Isiolo abattoir to delay as officials differ on who to equip the facility'. *The Press Point*, 10 August. https://thepresspoint.com/news-today/here-is-why-isiolo-abattoir-will-not-be-opened-soon/
Lunstrum, E. (2014). 'Green militarization: anti-poaching efforts and the spatial contours of Kruger National Park'. *Annals of the Association of American Geographers* 104(4), 816–832.
Mkutu, K. (2019). 'Pastoralists, politics and development projects: Understanding the layers of armed conflict in Isiolo County, Kenya'. *BICC Working Paper, 7/2019*. Bonn: Bonn International Center for Conversion (BICC). https://nbnresolving.org/urn:nbn:de:0168-ssoar-68074-7
Mkutu, K. (2020). 'Security dynamics in concervancies in Kenya: The case of Isiolo County'. https://nbn-resolving.org/urn:nbn:de:0168-ssoar-68256-7
Mkutu, K. and A. Boru (2019). *Rapid assessment of the institutional architecture for conflict mitigation*. Washington, DC: World Bank.

Mkutu, K., M. Müller-Koné, and E. A. Owino (2021). 'Future visions, present conflicts: The ethnicized politics of anticipation surrounding an infrastructure corridor in northern Kenya'. *Journal of Eastern African Studies* 15(4), 707–727. https://doi.org/10.1080/17531055.2021.1987700

Mkutu, K., C. Schetter, and E. A. Owino (n.d.). 'Arms and pastoralist conflict on the Kenya-Uganda border: Current trends and governance considerations'. *BICC Working Paper*. Bonn: Bonn International Center for Conversion (BICC).

Müller-Mahn, D., K. Mkutu, and E. Kioko (2021). 'Megaprojects – mega failures? The politics of aspiration and the transformation of rural Kenya'. *The European Journal of Development Research* 33(4), 1069–1090.

Okwany, C. C. O. (2016). *Kenya's foreign policy towards Somalia: A contribution to insecurity*. Ås: Norwegian University of Life Sciences (NMBU).

Okwany, C. C. O. (2020a). 'Countering violent extremism in the horn of Africa: How international interventions influence the Security of Civilians'. *Africa Amani Journal (AAJ)* 7(1).

Okwany, C. C. O. (2020b). 'The effect of Kenya's ontological (in-)security in the context of the Horn of Africa'. In *13th International Conference Egerton University*, Njoro, 25–27 March.

Okwany, C. C. O. (2020c). 'Kenya's foreign policy towards the Horn of Africa. A case of the Lamu Port-South Sudan Ethiopia Transport Corridor'. In *13th International Conference Egerton University*, Njoro, 25–27 March.

Okwany, C. C. O. (2023). *Territoriality as a method for understanding armed groups in Kenya and strengthening policy responses*. RESOLVE Network, United States Institute of Peace. https://doi.org/10.37805/pn2023.1.lpbi

Okwany, C. C. O., E. A. Owino, and Z. O. Sidha (2023). 'The nature of armed groups in Kenya in the context of territoriality'. In *Peace and security in Africa's borderlands*. Africa Amani Journal 9(2), 74–93.

Osamba, J. O. (2000). 'The sociology of insecurity: Cattle rustling and banditry in North-Western Kenya'. *African Journal on Conflict Resolution* 1(2), 11–37.

Owino, E. A. (2019). *The implications of large-scale infrastructure projects to the communities in Isiolo County: The case of Lamu Port South Sudan Ethiopia transport corridor*. Nairobi: United States International University-Africa.

Republic of Kenya (1965). *Sessional paper No. 10 of 1965 on African Socialism and its application to planning in Kenya*. Nairobi: Government Printers. https://repository.kippra.or.ke/handle/123456789/2345

Republic of Kenya (1986). *Sessional paper No. 01 of 1986 on economic management for renewed growth*. Nairobi: Republic of Kenya. https://repository.kippra.or.ke/handle/123456789/2679

Republic of Kenya (2006). *Ministry of livestock development and marketing report*. Nairobi: Government Printers.

The Round Table (1958). 'A constitution for Kenya: Mr. Lennox-Boyd's visit'. *Taylor and Francis* 48(190). https://doi.org/10.1080/00358535808452115

Sack, R. D. (1986). *Human Territoriality: Its Theory and History* (Vol. 7). London/ New York/ New Rochelle/ Melbourne/ Sydney: Cambridge University Press (CUP) Archive.

Schetter, C. and M. Müller-Koné (2021). 'Frontiers' violence: The interplay of state of exception, frontier habitus, and organized violence'. *Political Geography* 87, 102370. https://doi.org/10.1016/j.polgeo.2021.102370

Sharamo, R. (2014). 'The politics of pastoral violence: A case study of Isiolo County, Northern Kenya'. *Future Agricultures Consortium Working Paper* 95.

Veit, P. and K. Reytar (2017). *By the numbers: Indigenous and community land rights*. World Resource Institute. Retrieved 17.07.2021. www.wri.org/insights/numbers-indigenous-and-community-land-rights

White, B., S. M. Borras, Jr., R. Hall, I. Scoones, and W. Wolford (2013). 'The new enclosures: Critical perspectives on corporate land deals'. In *The new enclosures: Critical perspectives on corporate land deals*. Routledge, 13–42.

Wily, L. A. (2015). *Estimating national percentages of indigenous and community lands: Methods and findings for Africa*. Washington, DC: LandMark.

Media sources

07.05.2019. 'Isiolo Airport mosly idle despite sh2.7bn upgrade'. *Business Daily*.

31.07.2017. 'Isiolo International Airport: The Government of Kenya commissioned refurbished Isiolo Airport in July 2017. Located in Isiolo county, it is the country's fifth international airport.' *Airport Technology*.

17.05.2019. 'Spreading the net: Somali Islamists now target Kenyan recruits.' *Reuters*.

19.07.2020. 'Isiolo new abattoir to be operational in six months time'. *Daily Nation*.

04.07.2021. 'Completion of Isiolo export abattoir on course, official says.' *Daily Nation*.

15.10.2022. 'Kapedo deathtrap: A semi-territoriality advantage to the bandits'. *The Star*.

Index

aerotropolis 5, 9, 16, 65, 68, 89, 92, 120–134, 120–136; *see also* airport city
afforestation 29, 68, 72–75
airport city 98–99, 114–115; *see also* aerotropolis
airport expansion 1, 6, 12, 25, 57, 63, 70, 97–98, 102, 106–107, 110, 114–115, 146–147
airport land: concept 10–11
airport reserve: concept 10–11
airport wall 11, 103–106, 112–113
anticipation 8, 13, 22–42
aviation authority 48, 62, 65, 76, 82, 100, 138, 146–147

Cargo airport 8, 9, 10, 64, 99, 103–104, 114–115
China 25, 47, 58, 63, 72, 88, 145–146
citizenship 98, 102–103, 114–115
civil war 25, 102–103, 105, 115–116
colonial 2–3, 6, 8, 16, 45, 65–66, 68, 97, 99, 139, 142, 144–145, 151–152
compensation 10, 12–15, 26, 29, 34–35, 52, 59, 63, 67–70, 74–77, 88, 90–93, 102, 104, 108, 110, 113, 139, 146–148, 149, 152–153
connectivity 4, 8–10, 62, 64–65, 70, 73–75, 85, 87, 120, 122, 126, 130–133
corruption 6, 58, 142, 148
Côte d´Ivoire 87–119
customary tenure 12–13, 72, 100–102, 104–105, 107–110, 114, 142, 147–148

deforestation 27, 49, 51, 72, 75
demarcation 11, 97, 148–149
demolition 66, 105, 111, 112–115, 147
demonstration 17, 51, 59, 69, 113
destruction of habitat 6, 43, 45, 48, 51, 56, 71–73, 75, 132

destruction of livelihood 6, 52, 115, 147
disaster 51, 72, 84, 88, 111, 113
displacement 4–6, 13–14, 16, 66, 68, 73, 84–85, 89, 91–93, 97–98, 106, 111–112, 120, 123–125, 133, 139, 144, 146–147
dispossession 11, 12, 14, 34, 64, 66, 68, 75, 102, 104, 123–124, 126

economic corridor 63, 68, 125–126, 128, 130, 138, 145–147
elephants 43–61, 71–73
Environmental Impact Assessment 26, 45, 48, 68–70
Ethiopia 17, 88, 120, 138, 142
ethnopolitical conflict 137–159
Europe 1, 5, 8–9, 17, 143, 146
eviction 12, 17, 29, 97–98, 106, 109–115, 147, 153

farmers 4, 14–15, 49–58, 67–72, 75, 82, 88, 91, 93, 98, 103–107, 115, 121, 149
forest 65, 67, 71, 73

globalisation 9, 64, 121–122, 133
Global North 2, 5–9
Global South 2–9, 13
governance 4–5, 9–10, 12, 47, 69, 71, 74–75, 100, 108, 115, 120, 121, 127, 129, 131, 133, 144
grazing land 14, 66, 139, 141, 146, 149, 152
greenfield development 1, 6, 12, 14, 62, 67–71, 75, 124, 132

health 14, 17, 52, 121
human right violations 2, 6, 16, 97, 113
human-wildlife conflict 43–61

impoverishment 6, 84, 88, 93
India 26, 47, 51, 62–80, 81, 124, 133, 146

Indonesia 3, 7, 17, 81–96
informal settlement 1, 12, 14, 17, 111
infrastructure 3, 6–15, 22, 27, 30–31, 33, 55, 57, 62, 64–65, 68, 73–74, 85–88, 99, 122, 124, 126, 138, 142, 145–146
investment 4, 9–10, 25, 32, 37–38, 64–65, 74, 82, 87, 114, 120, 122–133

Kenya 120–136

land acquisition 28, 66–69, 101–102, 149; *see also* land appropriation
land appropriation 71–72, 75, 89, 100, 104, 110, 115, 146–147; *see also* land acquisition
land grabbing 49, 53–55, 57, 66, 72, 124, 142–145, 149
land prices 11, 14, 22, 32–33, 38, 54
land registration 101, 110, 142, 148
land speculation 32, 34, 83
Latin America 3, 17
litigation 3, 7, 14, 26, 37, 69, 71, 73, 104, 115, 147, 152–153

media 30, 32, 47, 50–51, 56, 59, 68, 97, 113, 115
military 8, 45, 98, 104–105, 140, 142, 149, 153
municipal policy 108, 110–111, 114–116, 120–136

narratives 11, 27–28, 31, 38, 51, 68, 70
national park 28, 30, 44–45, 49
Nepal 22–42
North America 1, 5, 8

pastoralism 137–159
petition 17, 56, 69
political participation 6, 7, 57, 89, 102, 151
privatisation 8–9, 64
promises 7, 11, 27–28, 30, 32, 34–36, 38, 50, 55, 70, 72, 75, 148
protest 1–2, 6, 13–14, 16–17, 37, 50–59, 67–73, 112–113, 115, 124

relocation 12–14; *see also* resettlement
resettlement 14, 22, 28, 34–35, 37, 68–69, 88, 98, 110, 146–147; *see also* relocation

Senegal 12, 13
South Africa 120–136
South Asia 3, 17
South-East Asia 3, 6, 25
special economic zone 5, 68, 120
Sri Lanka 43–61
suspension 23, 28, 30–36, 38, 68

Tanzania 14
temporality 27
tourism 4, 9–10, 22, 30, 36, 43, 45, 63–64, 71–72, 86, 93, 99, 114, 125, 130, 133

violence 16, 86, 120, 137
vision of development 4, 8, 14, 28, 68, 70, 73–74, 85, 87, 114, 122, 146, 151

water 4, 6, 14–15, 48, 50, 53–54, 57, 65, 71–72, 84, 103, 107, 113, 144
wildlife 26, 28, 43–61, 68, 70, 71–74

zoning 1, 11, 69, 71–72, 74, 97, 104, 112, 122, 139, 150